Computer-Aided Manufacturing and Design

Computer-Aided Manufacturing and Design

Editors

Qi Zhou
Seung-Kyum Choi
Recep M. Gorguluarslan

MDPI • Basel • Beijing • Wuhan • Barcelona • Belgrade • Manchester • Tokyo • Cluj • Tianjin

Editors
Qi Zhou
Huazhong University of Science
and Technology
China

Seung-Kyum Choi
Georgia Insitute of Technology
USA

Recep M. Gorguluarslan
TOBB University of Economics
and Technology
Turkey

Editorial Office
MDPI
St. Alban-Anlage 66
4052 Basel, Switzerland

This is a reprint of articles from the Special Issue published online in the open access journal *Applied Sciences* (ISSN 2076-3417) (available at: https://www.mdpi.com/journal/applsci/special_issues/Computer-Aided_Manufacturing_Design).

For citation purposes, cite each article independently as indicated on the article page online and as indicated below:

LastName, A.A.; LastName, B.B.; LastName, C.C. Article Title. *Journal Name* **Year**, *Article Number*, Page Range.

ISBN 978-3-03943-134-2 (Hbk)
ISBN 978-3-03943-135-9 (PDF)

Contents

About the Editors

Qi Zhou Ph.D., is an Associate Professor at the School of Aerospace Engineering, Huazhong University of Science and Technology (HUST), Wuhan, China. His research expertise areas include multi-fidelity surrogate model, airticial intelligence, and design optimization under uncertainty. He was awarded the National Defense Innovation Award, and obtained the Outstanding Young Researcher project from HUST. He was the session chair of the International Conference on System Modeling and Optimization committee. He has published over 80 peer-reviewed international journal and conference papers.

Seung-Kyum Choi began at Georgia Tech in Fall 2006. Dr. Choi's research interests include structural reliability, probabilistic mechanics, statistical approaches to the design of structural systems, multidisciplinary design optimization, and information engineering for complex engineered systems. He conducted a challenging research effort on the uncertainty quantification for the analytical certification of aerospace and underwater vehicles. The success of this challenging effort includes the development of state-of-the-art numerical techniques in statistical methods, and structural analysis methods with innovative uncertainty quantification techniques. Dr. Choi's current research focus is on the development of robust simulation and decision support tools to assist the management of complex engineered systems involving material structures design and multifunctional products. He is the principal author of a graduate-level textbook on the topics of probabilistic mechanics and reliability-based design optimization (Reliability-based Structural Design, Springer, 2007). He is an invited reviewer of numerous journals, including *Structural and Multidisciplinary Optimization, Probabilistic Engineering Mechanics, ASME JMD,* and *AIAA* journal. In addition, he has served as a Guest Editor for highly qualified journals, including *Journal of Engineering Design and Journal of Electronic Materials.*

Recep M. Gorguluarslan Ph.D., is an Assistant Professor at TOBB University of Economics and Technology in Turkey. He started his PhD in Mechanical Engineering at Georgia Institute of Technology in August 2011. He completed his PhD in December 2016. His PhD study was about developing a multi-level upscaling and validation approach for uncertainty quantification in additively manufactured lattice structures. In 2017, he worked at the Georgia Institute of Technology as a post-doctorate researcher in the project of developing an automated railway design approach of seats that can climb stairs, supported by ThsyennKrupp.

Editorial

Editorial for the Special Issue: Computer-Aided Manufacturing and Design

Qi Zhou [1], Seung-Kyum Choi [2,*] and Recep M. Gorguluarslan [3]

[1] School of Aerospace Engineering, Huazhong University of Science & Technology, Wuhan 430074, China; qizhouhust@gmail.com or qizhou@hust.edu.cn
[2] G. W. Woodruff School of Mechanical Engineering, Georgia Institute of Technology, 813 Ferst Drive, Atlanta, GA 30332, USA
[3] Department of Mechanical Engineering, TOBB University of Economics and Technology, 06560 Ankara, Turkey; rgorguluarslan@etu.edu.tr
* Correspondence: schoi@me.gatech.edu

Received: 10 July 2020; Accepted: 9 August 2020; Published: 14 August 2020

1. Introduction

Recent advancements in computer technology have allowed designers to have direct control over the production process through the help of computer-based tools, creating the possibility of completely integrated design and manufacturing processes. Over the last few decades, artificial intelligence (AI) techniques such as machine learning and deep learning have been topics of interest in computer-based design and manufacturing research fields. This Special Issue aims to collect novel articles covering artificial intelligence-based design, manufacturing, and data-driven design.

2. Content

This Special Issue comprises 10 selected papers that demonstrate the successful application of computer-based tools in design and manufacturing research fields.

Among these works, three papers focus on engineering optimization by combining computer-aided engineering (CAE) models with intelligent optimization algorithms. Specifically, in Reference [1], the finite element analysis (FEA) model for simulating the filling and packing stage was combined with a gradient-based algorithm and robust genetic algorithm to design the conformal cooling channels. In Reference [2], the hydraulic optimization of automotive electronic pumps was finished by combining the computational fluid dynamics (CFD) technology with a multi-island genetic algorithm. In Reference [3], the design optimization of an underwater vehicle base was successfully performed by integrating the FEA simulation-based design with the Kriging surrogate model and genetic algorithm.

Six of these papers focus on data-driven design and optimization. Specifically, in Reference [4], a stretchable micro-strip patch MSP (micro-strip patch) antenna-based strain sensor was optimized by a proposed design framework, which exploits dimensional reduction, machine learning-based surrogate modeling, structural optimization, and reliability assessment approaches. In Reference [5], a field repair kit for a complex product-service system was optimized in terms of the field inventory kit cost, while satisfying the availability requirement set by contract with the customer. In Reference [6], a methodology of a product image design integrated decision system based on Kansei engineering theory was developed. In Reference [7], to improve the quality of the large-scale assembly, an assemblability analysis and optimization method based on the coordination space model was developed. In Reference [8], a region-based convolutional neural network was constructed to recognize graphical symbols in piping and instrument diagrams. In Reference [9], the design specifications for a multifunctional console of Jangbogo class submarines that can accommodate, as much as possible, the anthropometric dimensions of Korean males were optimized.

The last paper [10] focuses on computer-based design for additive manufacturing. Specifically, the authors developed a design method to consolidate parts for considering maintenance and product recovery at the end-of-life stage.

3. Results

AI techniques shine in many areas, including the computer-based design and manufacturing research fields. The 10 papers described here show some successful applications of machine learning and intelligent optimization algorithms in different cases. It is believed that the collection of 10 papers in this Special Issue will be beneficial to readers who have interests in applying AI techniques in the computer-based design and manufacturing domain.

Author Contributions: All the Guest Editors contribute equally to the editorial paper of this Special Issue. All authors have read and agreed to the published version of the manuscript.

Funding: This editorial paper has been supported by the National Natural Science Foundation of China (NSFC) under Grant No. 51805179 and the National Defense Innovation Program under Grant No. 18-163-00-TS-004-033-01.

Acknowledgments: The Guest Editors sincerely thank all the authors for their excellent contributions to this Special Issue. Furthermore, we would like to thank all the anonymous reviewers for their selfless help in providing valuable comments and suggestions. Finally, the Guest Editors sincerely appreciate Lucia Li, the contact editor of this Special Issue, for her time and efforts.

Conflicts of Interest: The authors declare no conflict of interest.

References

1. Chung, C.-Y. Integrated Optimum Layout of Conformal Cooling Channels and Optimal Injection Molding Process Parameters for Optical Lenses. *Appl. Sci.* **2019**, *9*, 4341. [CrossRef]
2. Si, Q.; Lu, R.; Shen, C.; Xia, S.; Sheng, G.; Yuan, J. An Intelligent CFD-Based Optimization System for Fluid Machinery: Automotive Electronic Pump Case Application. *Appl. Sci.* **2020**, *10*, 366. [CrossRef]
3. Qian, J.; Yi, J.; Zhang, J.; Cheng, Y.; Liu, J. An Entropy Weight-Based Lower Confidence Bounding Optimization Approach for Engineering Product Design. *Appl. Sci.* **2020**, *10*, 3554. [CrossRef]
4. Hwang, S.; Gorguluarslan, R.; Choi, H.-J.; Choi, S.-K. Integration of Dimension Reduction and Uncertainty Quantification in Designing Stretchable Strain Gauge Sensor. *Appl. Sci.* **2020**, *10*, 643. [CrossRef]
5. Suh, E.S. Product Service System Availability Improvement through Field Repair Kit Optimization: A Case Study. *Appl. Sci.* **2019**, *9*, 4272. [CrossRef]
6. Xue, L.; Yi, X.; Zhang, Y. Research on Optimized Product Image Design Integrated Decision System Based on Kansei Engineering. *Appl. Sci.* **2020**, *10*, 1198. [CrossRef]
7. Cui, Z.; Du, F. A Coordination Space Model for Assemblability Analysis and Optimization during Measurement-Assisted Large-Scale Assembly. *Appl. Sci.* **2020**, *10*, 3331. [CrossRef]
8. Yun, D.-Y.; Seo, S.-K.; Zahid, U.; Lee, C.-J. Deep Neural Network for Automatic Image Recognition of Engineering Diagrams. *Appl. Sci.* **2020**, *10*, 4005. [CrossRef]
9. Lee, J.; Cho, N.; Yun, M.-H.; Lee, Y. Data-Driven Design Solution of a Mismatch Problem between the Specifications of the Multi-Function Console in a Jangbogo Class Submarine and the Anthropometric Dimensions of South Koreans Users. *Appl. Sci.* **2020**, *10*, 415. [CrossRef]
10. Kim, S.; Moon, S.K. A Part Consolidation Design Method for Additive Manufacturing based on Product Disassembly Complexity. *Appl. Sci.* **2020**, *10*, 1100. [CrossRef]

Article

An Intelligent CFD-Based Optimization System for Fluid Machinery: Automotive Electronic Pump Case Application

Qiaorui Si [1], Rong Lu [1], Chunhao Shen [1], Shuijing Xia [2], Guochen Sheng [1] and Jianping Yuan [1,*]

[1]   National Research Center of Pumps, Jiangsu University, Zhenjiang 212013, China; siqiaorui@ujs.edu.cn (Q.S.); lurongdemail@163.com (R.L.); 18852854933@163.com (C.S.); shengguochenujs@163.com (G.S.)
[2]   Pierburg Huayu Pump Technology Co. Ltd., Shanghai 201900, China; Shuijing.Xia@phptc.com
*   Correspondence: yh@ujs.edu.cn; Tel.: +86-139-5280-1593

Received: 12 November 2019; Accepted: 31 December 2019; Published: 3 January 2020

Abstract: Improving the efficiency of fluid machinery is an eternal topic, and the development of computational fluid dynamics (CFD) technology provides an opportunity to achieve optimal design in limited time. A multi-objective design process based on CFD and an intelligent optimization method is proposed in this study to improve the energy transfer efficiency, using the application of an automotive electronic pump as an example. Firstly, the three-dimensional CFD analysis of the prototype is carried out to understand the flow loss mechanism inside the pump and establish the numerical prediction model of pump performance. Secondly, an automatic optimization platform including fluid domain modeling, meshing, solving, post-processing, and design of experiment (DOE) is built based on three-dimensional parametric design method. Then, orthogonal experimental design and the multi-island genetic algorithm (MIGA) are utilized to drive the platform for improving the efficiency of the pump at three operating flowrates. Finally, the optimal impeller geometries are obtained within the limited 375 h and manufactured into a prototype for verification test. The results show that the highest efficiency of the pump increased by 4.2%, which verify the effectiveness of the proposed method. Overall, the flow field has been improved significantly after optimization, which is the fundamental reason for performance improvement.

Keywords: automatic design; intelligent optimization method; CFD; fluid machinery; pumps

---

1. Introduction

Fluid machinery is an important energy conversion device, which is widely used in important sectors of the national economy such as hydropower, chemical processes, automobiles, nuclear power, and national defense [1,2]. With the deepening of energy saving and emission reduction, it is very important to improve the conveying efficiency of fluid machinery. Take the automobile industry as an example, the design of efficient cooling systems which are driven by blade pumps play an important role in the development of new energy vehicles. The former mechanical cooling water pump is mainly driven and coupled with the engine speed, which may either overcool or undercool. Electric water pumps are powered by adjustable speed motors and regulate operating conditions according to the cooling needs [3–5]. The pump unit coupled with the control program not only minimize the output power, but also meets the needs of the electric and intelligent development of the automotive industry. However, the efficiency of the pump is often below 40% due to unreasonable hydraulic design, leaving space for considerable energy savings. Moreover, pump geometries are mostly calculated based on ideal flow theory under a single flow rate using the traditional one-dimensional design method. However, the pump usually deviates from the optimal working condition, resulting in unstable flow phenomena such as secondary flow and flow separation inside the flow channel during the actual operation [6–8]. Thus, the study on the multi-objective optimization method of efficient pumps is also very important to the designers. However, there is still a lack of theoretical support and effective optimization tools for pump designing in the background of faster product updates.

Pump design theory has evolved from traditional one-dimensional design to three-dimensional design. With the rapid development of computer technology, the application of computational fluid dynamics (CFD) combined with optimization methods have become a popular and effective technique in turbomachinery design [9]. Wu et al. [10] investigated the effects of trailing edge modification on the performance of the mixed-flow pump through CFD analysis. Zhou et al. [11] compared the internal flow characteristics of a new kind of three-dimensional surface return diffuser to traditional ones using steady CFD simulation in order to improve the hydrodynamic performance of the deep-well centrifugal pump. Osman et al. [12] numerically investigated two multistage axially split centrifugal pumps with different channel designs between its stages, the flow losses were compared by entropy production. Wang et al. [13] established an energy loss model (ELM) to determine the relationships among the different types of energy losses in a multistage centrifugal pump, and a method was proposed to optimize the pump efficiency based on the ELM and CFD. The optimal design method of pumps has been studied extensively by many scholars. There are several methods to improve the pump performance, such as the empirical design method, approximate model method, and optimization algorithm. Liu et al. [14] implemented the orthogonal design with five factors and four levels to optimize a multiphase pump, the influence of each of the factors on the pressure rise was estimated, and the optimized ranges of these parameters were determined. The above method achieved some success but could not overcome the influence of human factors on the optimization results because it still relies on the empirical coefficients. In recent years, intelligent optimization algorithms have been developed rapidly, such as the genetic algorithm, ant colony algorithm, no-no search algorithm, simulated annealing algorithm, particle swarm algorithm, and so on [15–19]. With the optimization of pump performance, these intelligent algorithms are gradually adopted by pump researchers. Pei et al. [20] constructed an accurate nonlinear function between the optimization target and the pump design variables by utilizing an artificial neural network, and the modified particle swarm algorithm was further applied to optimize the mathematical model. Yuan et al. [21] adopted optimal Latin hypercube design, CFD simulation combined with the Kriging model were used to achieve the sample points for space-filling and establish the approximate optimization model, the best combination of impeller parameters were finally obtained by solving the approximation model with a genetic algorithm. Zhang et al. [22] proposed a multi-objective method to optimize a double suction centrifugal pump based on the Simulation Kriging Experiment (SKE) approach. Wang et al. [23] used radial basis function neural network combined with the Non-dominated Sorting Genetic Algorithm-II (NSGA-II) genetic algorithm to optimize a high-speed, mixed-flow pump. However, the optimization periods of these two methods are relatively short and it is difficult to obtain an overall optimal solution. In order to eliminate the deviation generated by the build of the approximation model, the researchers used the natural heuristic algorithm to optimize the performance globally. For example, Zhang et al. [24] utilized a genetic algorithm to optimize the geometry of a spiral axial-flow mixing pump. In recent years, a lot of intelligence algorithms have been proposed. Because there is no crossover and mutation process in the genetic algorithm, the convergence speed is greatly accelerated [25,26]. In summary, the pump optimization methods have undergone an improvement process from single target to multi-objective optimization, from experience to theory. Currently, with the advance of computer technology, automation, and artificial intelligence, automatic design methods without human intervention are being developed and applied in various fields.

This study focuses on the hydraulic optimization of automotive electronic pumps. Firstly, numerical simulations for a given model were carried out to build the prediction process of pump efficiency. Secondly, the target of this multi-objective optimization was established by weighting analysis. Then, an automatic optimization platform was established including parametric design, CFD and DOE. Orthogonal design, a kind of DOE method, was implemented in advance to estimate the influence of four selected factors on pump efficiency and head. The optimized ranges of these parameters were also determined. After, one kind of DOE namely the multi-island genetic algorithm (MIGA) was utilized to drive the platform for optimizing the weighted average efficiency under three operating flowrates. Finally, the optimized hydraulic scheme was obtained in a limited 375 h by using a normal desktop computer without running in parallel and was proven to improve the pump efficiency by 4.2% through the experiment. The proposed method may provide some theoretical guidance for hydraulic optimization of fluid machinery.

2. Optimization Design Process and Modeling Method

2.1. Optimization Process

In traditional design of fluid machinery, taking pumps for example, research procedures involve manufacturing prototypes, performance tests, results analyses, and optimal design. The design is mostly two-dimensional and relies on the experience of the designer. However, creating prototypes may cost a lot of time and money. Moreover, a large number of prototype tests would inevitably cause significant manufacturing and testing errors. The automatic design with the intelligent optimization method of fluid machinery presented in this study is based on the combination of CFD technology and optimization theory. An accurate CFD simulation method needs to be established first, which could provide virtual experimental results for optimization. Meanwhile, an automatic optimization platform including all CFD process such as fluid domain modeling, meshing, solving, post-processing, and design of experiment (DOE) was built for future optimization processes. The modeling of the pump geometry is based on the three-dimensional parametric design method, which means the fluid domain could be modified by DOE arrangements. The main parameters that influence the hydraulic performance of centrifugal pumps were screened out, and orthogonal design was applied to obtain the spacing-filling sample points. Significance test of optimization variables was conducted based on analysis of variance (ANOVA). Further, the MIGA optimization method was utilized to drive the platform for improving the efficiency of the pump automatically under three flowrates. Finally, an optimized scheme was obtained and the prototype was 3D printed to experimentally validate the optimal results. The whole optimization process is shown in Figure 1.

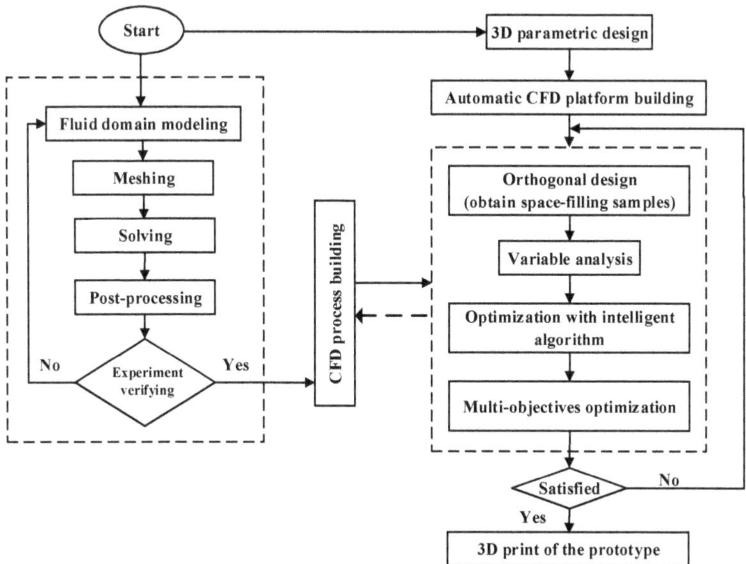

Figure 1. The flowchart of optimization process. CFD, computational fluid dynamics.

2.2. Pump Model

The electronic pumps used in a car are always single-stage, single-suction centrifugal pumps equipped with a Direct Current (DC) brushless motor. They are characterized by simple structure, reliable operation, and convenient speed regulation. The motor contains the stator, the rotor, and shaft. The rotor will rotate after power-on under electromagnetic induction. Moreover, the electronic pump is equipped with a control unit that is connected to the driving computer. The rotational speed of the pump would be adjusted by pulse width modulation (PWM) controls according to heat dissipation of the cooling system, which may ensure the pump operates at a constant operation area after dimensionless processing [1].

2.2.1. Pump Geometry and Losses Analysis

The structure of automotive electronic pumps used in this study is shown in Figure 2a. The coolant liquid is sucked into the impeller, obtains pressure head after rotating with the impeller, and enters into the volute passage under centrifugal force. In this process, electrical energy is converted into the pressure energy of the fluid with a lot of losses, such as mechanical loss, leakage loss, disc friction losses, and hydraulic loss. Mechanical loss refers to the energy loss due to the mechanical friction in the pump, including the disk friction loss, bearing loss, and shaft seal loss. Leakage loss refers to the clearance leakage in the pump including the leakage loss at the front ring, rear ring, and balance holes. Hydraulic loss refers to the energy loss through the flow passage, including the inlet section, outlet section, pump cavity, impeller, and volute. The flow inside the pump passage is full 3D unsteady turbulence flow, including flow separation, backflow, circulation, instability flow, jet-wake flow, vortex, and even cavitation. As the most important part of energy transfer, the geometry of the impeller has a significant effect on hydraulic losses, pumping ability, and inherent reliability. Impeller-related losses dominate the pump efficiency. Therefore, the optimization of this study is mainly aimed at the impeller.

The pump performance characteristic curves at different rotational speeds are shown in Figure 2b. The pump would work at point A in actual operation at design rotational speed n_d because of system resistance. Similarly, the pump would work at point A_1 in actual operation at rotational speed n_1 based on similarity law if Reynolds number changes within a certain range. The related losses of the pump under different rotational speeds also follows the similar law. Thus, it is equivalent to evaluate the pump performance only at the design rotational speed. The hydraulic parameters of the selected model pump are as follows: design rotational speed n_d is 5400 r/min, design head H_d is 7.5 m, design flow rate Q_d is 1.4 m^3/h. Table 1 provides a list of the main geometric parameters of the pump. Blade angle is relative to tangential direction. Considering the inlet pressure of the pump in the real system, this study does not involve cavitation problems.

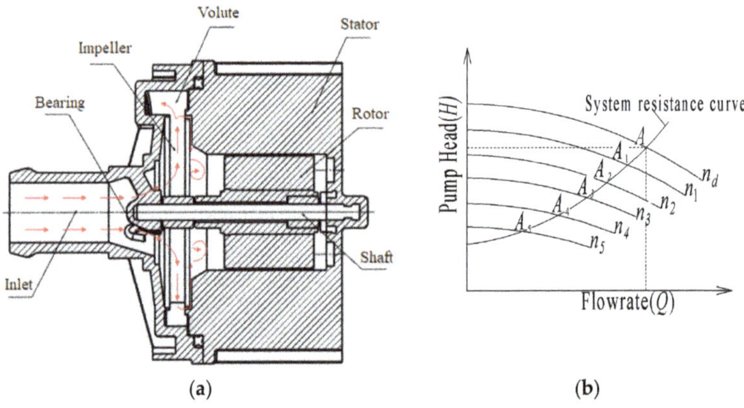

Figure 2. Automobile electronic pump model: (**a**) Structure diagram; (**b**) Performance characteristic curves.

Table 1. Main geometric parameters of the pump.

Description	Parameter	Symbol	Unit	Value
Impeller	Inlet diameter	D_1	mm	20
	Outlet diameter	D_2	mm	47
	Outlet width	b_2	mm	2.2
	Blade inlet angle	β_1	°	20
	Blade number	Z	-	5
	Blade outlet angle	β_2	°	41
Volute	Inlet diameter	D_3	mm	50
	Inlet width	b_3	mm	6
	Pipe diameter	D	mm	17

2.2.2. Optimization Objectives

The objective of the optimization is to build the relationships between the geometrical parameters of the impeller and the various kinds of energy losses in the pump, with the ultimate goal of minimizing the total energy loss. Thus, the following geometric parameters of impeller were selected as optimization variables based on the results of previous practice, such as the impeller outlet diameter D_2, blade outlet width b_2, blade number Z, blade outlet angle β_2, blade inlet angle β_1, leading edge tangential angle t_3, trailing edge tangential angle ϕ, and blade thickness δ. Moreover, the electronic water pump needs to operate under multiple working flowrates according to the working state of the cooling system. Therefore, a multi-point optimization is necessary. To broaden the high efficiency range of the model pump, the weighted average efficiency of three flow conditions, namely $0.8Q_d$, $1.0Q_d$, $1.25Q_d$, were set as the objective function, and the design head was set as a constraint.

In general, the time of optimization with the intelligent method will be increased with both the number of parameters and the range of each parameter. In this study, exactly four parameters, impeller outlet diameter D_2, blade outlet width b_2, blade number Z, blade outlet angle β_2, were selected for the orthogonal experiment at the first step. These four parameters are considered to play most important roles in the comprehensive performance of centrifugal pumps [6,13]. Local optimization can be achieved, and then the range of parameters can be reduced in this step. However, because the optimization is also a multi-objective problem, it is necessary to establish the model with the highest weighted average efficiency under three working flowrates. The optimization problem can be formulated as:

Solve $X = [D_2, b_2, Z, \beta_2]^{\mathrm{T}}$ make:

$$f(X) = \frac{\sum\limits_{i=1}^{3} \omega_i \cdot \eta_i(X)}{\sum\limits_{i=1}^{3} \omega_i} \rightarrow \max, \tag{1}$$

$$\text{subject to} \quad \begin{matrix} X \in R \\ H_d \geq 7.5\text{m} \end{matrix}, \tag{2}$$

where X is a vector of all design variables, R is the set of variable ranges, η_1, η_2, η_3 represent the efficiency of $0.8Q_d$, $1.0Q_d$, and $1.25Q_d$, respectively. Accordingly, ω_i represents the weighting factors.

Analytic hierarchy process (AHP) was developed in the late 1970s by Saaty [27]. Due to its simplicity, ease of use, and great flexibility, this technique has been widely used as a decision model to deal with multi-criteria evaluation. However, AHP does not take into account the inherent uncertainty and imprecision. Moreover, the comparison matrices used in AHP are combined with crisp scales. To deal with the uncertainty or vagueness of data, fuzzy analytic hierarchy process (FAHP) was utilized to derive the weight of every objective [28,29]. In FAHP, the importance of factor A to factor B is obtained and the fuzzy judgment matrix is generated after comparing different factors. This study use FAHP with the following steps.

Firstly, build and construct a fuzzy complementary judgment matrix. The pairwise comparison among factors is made to determine the importance of one factor relative to the other factors, and then the judgment matrix is generated:

$$A = \left(a_{ij}\right)_{N \times N},$$ (3)

where a_{ij} represents the evaluation of the relative importance of the factor i to the factor j, $a_{ji} = 1/a_{ij}$.

Secondly, calculate weight vector:

$$\omega_i^* = (b_1, b_2, \ldots, b_N)^{\mathrm{T}},$$ (4)

where $i = 1, 2, 3, \ldots, N$; $b_i = \sqrt[N]{a_{i1} \cdot a_{i2} \cdots a_{iN}}$

Thirdly, calculate the normalized weight value for each weight vector obtained from Equation (4):

$$\omega_i = b_i / \sum_{i=1}^{N} b_i.$$ (5)

Fourthly, sum the elements in each column:

$$S_i = \left(\sum_{j=1}^{N} a_{1j}, \sum_{j=1}^{N} a_{2j}, \ldots \sum_{j=1}^{N} a_{Nj} \right)^{\mathrm{T}}.$$ (6)

Fifthly, calculate the maximum eigenvalue of the matrix. It would be used to calculate the following consistency index.

$$\lambda_{\max} = \sum_{i=1}^{N} \Omega_i \cdot S_i,$$ (7)

where $\Omega_i = (\omega_1, \omega_2, \ldots, \omega_N)$

Sixthly, consistency test. A consistency ratio (CR) is obtained in addition to the corresponding principal eigenvector when using FAHP for criteria weighting. It represents the priority vector which is integrated by the intended weights. The consistency ratio is used to measure if this eigenvector estimates the weight vector well. It is obtained by division of the consistency index (CI) and the appropriate average random consistency index (RI):

$$CR = \frac{CI}{RI},$$ (8)

$$CI = \frac{\lambda_{\max} - N}{N - 1},$$ (9)

where $\lambda_{\max}$ is the maximum eigenvalue of the comparison matrix, and N is the order of objective matrix. The value of the RI is related to N, which can be obtained from Table 2. If CR is greater than 0.1, the comparison matrix is inconsistent and should be revised.

Table 2. Random consistency index (*RI*).

N	1	2	3	4	5	6	7	8	9	10	11
RI	0	0	0.58	0.9	1.12	1.24	1.34	1.41	1.45	1.49	1.51

The second step for the optimization is to use the intelligent method. The impeller parameters mentioned before, the impeller outlet diameter D_2, blade outlet width b_2, blade number Z, blade outlet angle β_2, blade inlet angle β_1, leading edge tangential angle t_3, trailing edge tangential angle ϕ, and blade thickness δ, are used for variable parameters to find the optimal combination at this step. Batch command is used to drive the following cycle of the numerical simulation process.

2.3. Numerical Method

The key to intelligent optimization is to build a numerical calculation platform which runs automatically. So, the reliability of the numerical calculation process is particularly important.

2.3.1. Establishing the Calculation Domain

The hydraulic performance of the model pump is predicted by means of a CFD process in this study. The computational domain of the original pump is shown in Figure 3, which is comprised of four components, namely the inlet, impeller, pump chamber, and volute. In order to consider the full development of turbulence, the inlet and volute outlet pipes are properly extended. The inlet domain is directly connected to the impeller domain with a rotor-stator interface. The impeller domain is also directly connected to the chamber domain with a rotor-stator interface.

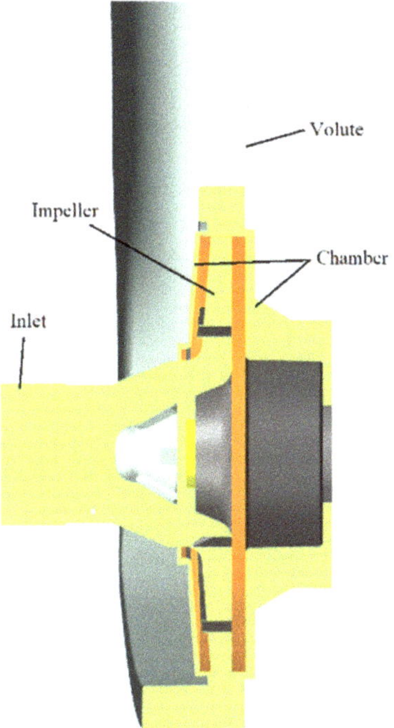

Figure 3. Model of computational domains.

2.3.2. Mesh Sensitivity Analysis

The calculation domains should be discretized by meshes before simulation. Due to the strong adaptability to complex geometry, tetrahedral unstructured grids generated by software Ansys ICEM14.1 are adopt in this study. The meshes of all computational flow domains are displayed in Figure 4.

Figure 4. Computational meshes.

The number of mesh elements can significantly influence the simulation results. Seven sets of meshes are used for grid independence analysis, with elements rising from 34×10^4 to 251×10^4. Grid convergence study is done at design flowrate using the Reynolds-averaged Navier-Stokes equations (RANS) method and boundary conditions described below. The Reynolds number value is about 6.2×10^5 based on the formula $R_e = u_2 \cdot D_2/2v$. u_2 represents the tangential velocity of the flow at the blade tip clearance under rated rotational speed 5400 r/min. v represents the kinematic viscosity of the fluid. So, it is reasonable to perform the following simulations in the framework of RANS. Seen from Figure 5, the overall difference of pump head is within 1% when the number of mesh elements exceed 1.68 million, which means the grid number has little effect on the calculation results. In order to balance the computational accuracy and the total calculation time, meshes with 168 million elements are employed for the following investigation.

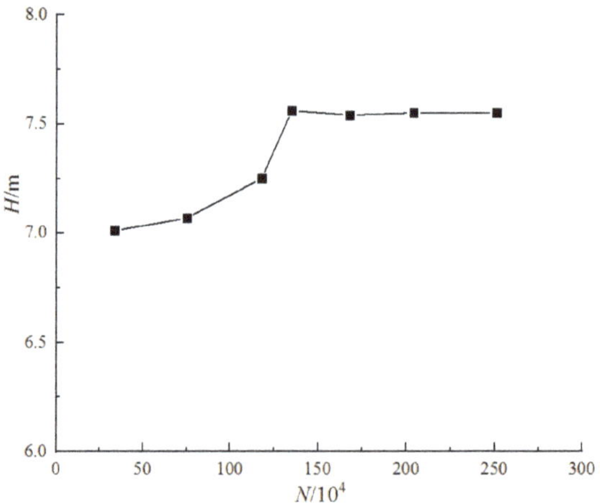

Figure 5. Analysis of grid independence.

2.3.3. Governing Equations and Boundary Conditions

The three-dimensional turbulent flow inside the pump is basically controlled by the law of mass and momentum conservation. For steady and incompressible flow, the control equations can be formulated as:

$$\nabla \cdot \vec{V} = 0 \tag{10}$$

$$\frac{d\vec{V}}{dt} = \vec{F} - \frac{1}{\rho}\nabla p + v\nabla^2\vec{V}, \tag{11}$$

where $\vec{V}$ represents the velocity vector, p represents pressure, $\vec{F}$ represents mass force, v represents kinematic viscosity, ρ represents the fluid density.

Due to different turbulence models influencing the calculation results of specific flow problems, selecting the suitable turbulence model can improve the calculation accuracy. In the present study, the fluid medium is set as water at 25 °C. The simulation is conducted by ANSYS CFX14.1 solver, which provides a number of turbulence models. Among the turbulence models, k-ε and k-ω are known in engineering applications. The ε equation of the standard k-ε model must use a wall function to solve the terms, and the simulation results are not accurate in the case of severe flow separation. The standard k-ω model is improved for a low Reynolds number and shear flow. Moreover, the Shear Stress Transfer (SST) k-ω turbulence model uses a hybrid function which acts as a standard k-ω model on the near wall and the standard k-ε model on the far wall area to correct the turbulent viscosity formula, while taking into account the turbulent shear stress. It has been proven to be more suitable for rotating machinery because it provides a better solution to the boundary layer [30]. The internal flow of the rotating pumps is three-dimensional turbulence flow, with strong flow separation, rotor-stator interaction, backflow, and so on. In addition, the electronic water pump has a high speed, and its flow losses mainly appear near the wall surface. Therefore, the SST k-ω turbulence model is adopted in this study to simulate the internal flow of the model pump based on the finite volume method.

The boundary conditions consist of an imposed stable total pressure with a turbulence intensity of 5% at the inlet and flow rates at outlet. All physical surfaces are set as no-slip walls. Considering the rotational characteristic, the rotating coordinate system is applied in impeller domain. The interfaces between the rotational domain and stationary ones are set as frozen rotor for steady simulation. The root mean square (RMS) residual error is used to judge whether the calculation is converged, setting values as 10^{-5}.

2.3.4. Experiment Validation

According to the Bernoulli equation, the pump head H is formulated as:

$$H = \frac{p_2 - p_1}{\rho g} + \frac{v_2{}^2 - v_1{}^2}{2g} + (z_2 - z_1), \tag{12}$$

where p_1 and p_2 imply the inlet static pressure and outlet static pressure respectively; v_1 and v_2 denote the average velocities of the inlet and outlet section respectively; z_1 and z_2 are the heights in vertical direction at the inlet and outlet of the model pump; ρ denotes the fluid density, 997 kg/m^3; g denotes the local acceleration of gravity, 9.8 m/s^2.

The efficiency of centrifugal pump is the ratio of useful power to input power, defined as follows:

$$\eta = \frac{\rho g Q H}{P_e}, \tag{13}$$

where P_e denotes the shaft power calculated by the input power and motor efficiency. The way to obtain the value of P_e in an experiment is different from the simulation because the torque T acting on the rotor is easy to obtain in CFD. P_e is equal to $n_{\mathrm{d}} \cdot T / 9550$.

In order to validate the accuracy of the CFD results, a test rig as shown in Figure 6 was set up to measure the pump performance. The model pump is connected to the rigid pipes by two rubber hoses. The flow rate is obtained by a LWGY-MIK-DN20 liquid turbine flowmeter with a range of 0.8–8 m^3/h and accuracy of 0.5%. The head of the pump is measured by a CYG1204F type differential pressure transmitter with an uncertainty of 0.1% and range of 0–200 kPa. The PSW 30–36 programmable DC power is used to drive the pump. The output voltage range of the power is 0–30 V, and the maximum output current is 36 A. The data acquisition system consists of LabVIEW acquisition program and NI6343 data acquisition card. The estimated uncertainty of head and efficiency measurement is below 0.32%, and the random uncertainty is no more than ±0.1%.

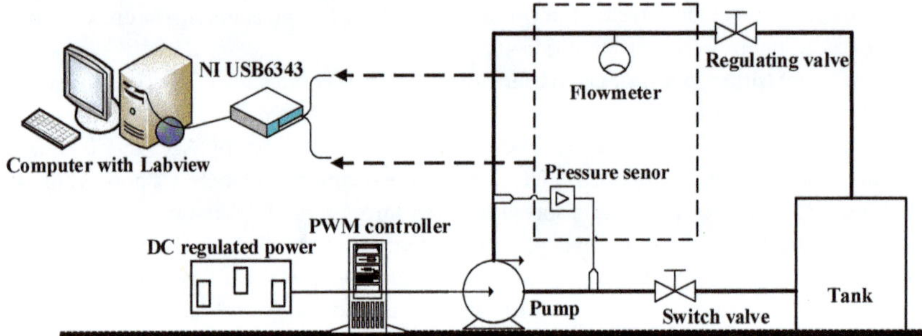

Figure 6. The schematic diagram of test rig.

The obtained pump hydraulic performance curves from CFD and experiment are compared, as shown in Figure 7. The pump efficiency obtained from the test has been converted into the hydraulic one by removing motor efficiency, bearing efficiency, and leakage efficiency. As seen from the results, the numerical ones are always consistent with the changing trends of the experimental curves. Both head and hydraulic efficiency of the simulation results are higher than the experimental values, which might be because the material roughness is not considered in the numerical simulation. The deviations of head at $0.8Q_d$, $1.0Q_d$, $1.25Q_d$ are 4.9%, 4.3%, 1.7%, respectively. The deviations of efficiency at $0.8Q_d$, $1.0Q_d$, $1.25Q_d$ are 3.6%, 4.9%, 4.8%, respectively. Therefore, it can be stated that the employed numerical method is reliable for the optimization at three selected flowrates.

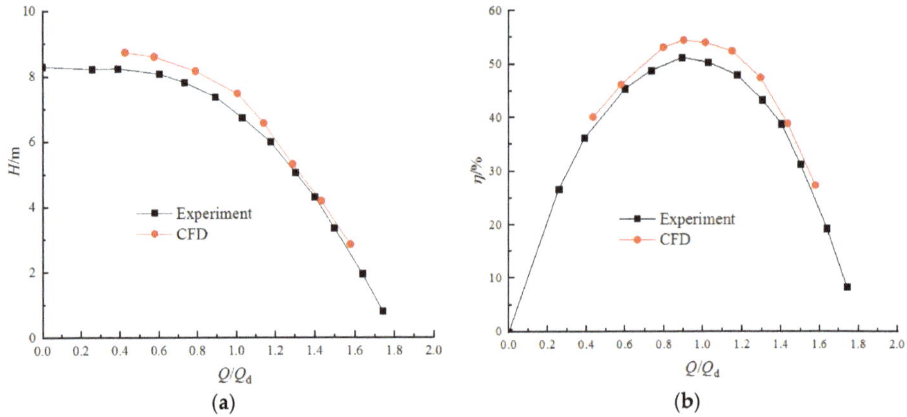

Figure 7. Comparison of pump performance between numerical and experimental results: (**a**) Head; (**b**) Efficiency.

3. Automatic Intelligent Optimization

3.1. Construction of Automatic Optimization Platform

In order to reduce manual intervention and save optimization time, an automatic optimization platform for CFD is established. The platform integrates a 3D parametric design module, mesh division module, pre-processing module, solver, and post-processing module with computer batch processing commands, which can automatically realize the combination of parametric design, grid generation, pre-process for CFD, simulation, and post-process for CFD. The operation process of this optimization platform is shown in Figure 8. The numerical optimization algorithm scheme is set in the DOE part.

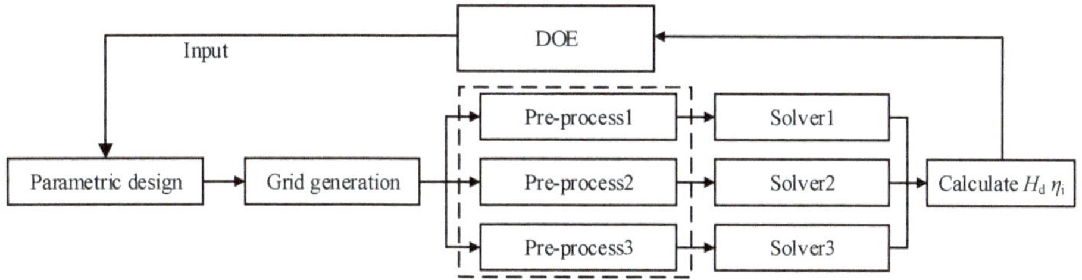

Figure 8. Operation process of the automatic optimization platform.

A computer-aided optimization platform, namely Isight, has been widely used in multidisciplinary design and optimization. Its rich component library and algorithm library, visualization of the running process, and powerful data analysis functions are useful for optimization design in many fields [31]. This study uses this software to realize the integration between 3D modeling software CFturbo and numerical simulation software Ansys CFX14.1. Thus, an intelligent optimization platform for multi-conditions of automotive electronic water pumps is built, as shown in Figure 9.

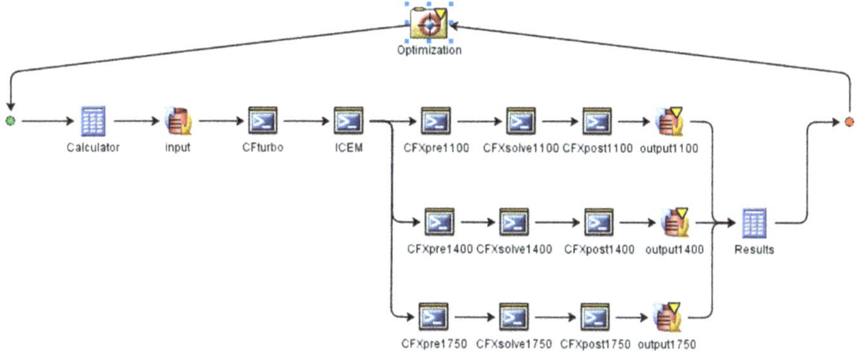

Figure 9. Integrated optimization platform of automotive electronic water pump.

3.2. Orthogonal Design Optimization

DOE is one of the most important methods in today's product development and performance optimization. Its function is similar to the mathematical arrangement, which makes a large number of data reasonably and orderly arranged. It provides a scientific experimental scheme for designers. The two main concepts in the experimental design are factors (input variables in the design) and levels (the number of values in each variable). In this study, three levels are selected for each factor, as shown in Table 3.

Table 3. Levels of factors in orthogonal experiments.

Levels	Factors			
	A D_2/mm	**B** b_2/mm	**C** Z	**D** β_2/(°)
1	44	3	4	30
2	45	3.2	5	33
3	46	3.5	6	35

As is listed in Table 4, a pair-wise comparison square matrix was made from the comparison factors based on experience, the diagonal elements of the matrix are always 1 because the same factors have the same weightage. The weight value (ω_i) of each objective is then calculated based on Equation (8). As seen from Table 4, CR is less than 0.1, which means the values of ω_i listed in the last column can be applied in this multi-objective problem.

Table 4. Evaluation results and weight factors. CR, consistency ratio.

Objective	η_1	η_2	η_3	ω_i
η_1	1	2/3	10/11	0.2786
η_2	3/2	1	5/4	0.4059
η_3	11/10	4/5	1	0.3155
CR		0.000725		

Theoretically, there are 81 (calculated by 3^4) kinds of schemes, while nine kinds use orthogonal arrays to have the same effect. Orthogonal arrays are compiled into a standardized table using the mathematical "orthogonality" principle to arrange the experimental program scientifically. This method was first applied to engineering design by Plackett and Burman [32]. A notational scheme to characterize the orthogonal table, as Equation (14) represents:

$$L_m\left(n^p\right), \tag{14}$$

where, n is the number of levels; p is the number of factors; m is the number of schemes.

In this study, L_9 (3^4) orthogonal table is applied to obtain the space-filling samples. Four factors and three levels are reasonably divided into nine groups of experimental schemes. The evaluation indexes results $f(X)$ are shown in Table 5, which are calculated according to Equation (1) based on the CFD method.

Table 5. Test schemes.

Number	Levels				Corresponding Parameters				Results
	A	B	C	D	D_2	b_2	Z	β_2	$f(X)$
1	A_1	B_1	C_1	D_1	44	3	4	30	46.11
2	A_1	B_2	C_2	D_2	44	3.2	5	33	50.92
3	A_1	B_3	C_3	D_3	44	3.5	6	35	50.81
4	A_2	B_1	C_2	D_3	45	3	5	35	50.89
5	A_2	B_2	C_3	D_1	45	3.2	6	30	51.21
6	A_2	B_3	C_1	D_2	45	3.5	4	33	48.35
7	A_3	B_1	C_3	D_2	46	3	6	33	51.53
8	A_3	B_2	C_1	D_3	46	3.2	4	35	49.73
9	A_3	B_3	C_2	D_1	46	3.5	5	30	51.64

3.2.1. Range and Sensitivity Analysis

The range analysis was performed to evaluate the influence of different factors on the evaluation index. The factor with the largest range was considered as the most sensitive factor, which had the greatest impact on the evaluation index. First of all, the comprehensive average value of t at the same level of each factor was obtained by Equation (15) On this basis, the range of each influencing factor was obtained by Equation (16), subtracting the minimum value from the maximum value of t at different levels of each factors. The results are shown in Table 6.

$$t_i = \frac{T_i}{r}, \tag{15}$$

$$R = \max(t_1, t_2, t_3) - \min(t_1, t_2, t_3), \tag{16}$$

where T_i represents the sum of all test target values at the i level; r is the number of different factor levels; t_i denotes the average of the test target values and R is the range.

Table 6. Range analysis.

Target	*f(X)* %			
	A	**B**	**C**	**D**
t_1	49.28	49.51	48.06	49.65
t_2	50.15	50.62	51.15	50.27
t_3	50.97	50.27	51.18	50.48
R	1.69	1.11	3.12	0.83

In order to visually reflect the impact size and trend of each factor on the evaluation target, the average value and factor level of the test target are drawn as a factors and indicators trend graph, as is shown in Figure 10. As seen from the range value analysis, the order of importance on efficiency is C, A, B, and D. The primary and secondary order of influence of each factor on the target is: $Z > D_2 > b_2 > \beta_2$. The blade number Z has the greatest impact on pump efficiency, followed by the factor A. As the blade number Z and impeller outlet diameter D_2 increase, the hydraulic performance improves significantly. Generally, the effect of impeller outlet width b_2 on efficiency is relatively small.

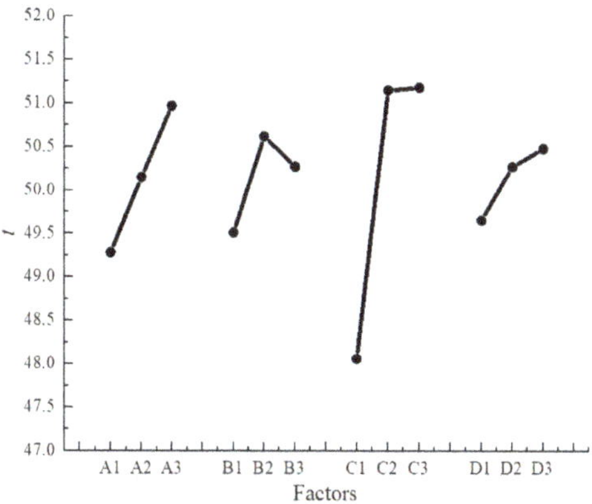

Figure 10. Range analysis diagram.

Analysis of variance (ANOVA) is an effective method to determine the significance of controllable factors on the research results [33]. It tests the variables by mean squaring and estimates experimental errors at specific levels. *F*-test is always used in ANOVA to analyze whether a particular design has any significant change in quality standards. In analyzing the parameters, the *F*-test tool based on ratio of mean square and residual error is used to find the significance of a factor. It is evaluated by using the equations below.

The sum of squared total deviation SS_T and total degrees of freedom f_T are defined as:

$$SS_T = \sum_{j=1}^{m} f_j^2(x) - \frac{T^2}{m} (j = 1, 2, \ldots, m)$$
$$f_T = m - 1$$

(17)

where m is the number of simples, $m = 9$ in this study. T is the sum of the nine scheme results. The sum of square SS_k and degrees of freedom of each factor f_k are defined as follows:

$$SS_k = \frac{1}{n} \sum_{k=1}^{n} T_k^2 - \frac{T^2}{m} (k = 1, 2, \ldots, n)$$
$$f_k = n - 1$$

(18)

where $k = 1, 2, 3, 4$ represents the factors A, B, C, D, respectively, $n = 4$.

The sum of the square SS_E and degree of freedom of the deviations f_E are defined as follows:

$$SS_E = SS_T - \sum_{k=1}^{n} SS_k$$
$$f_E = f_T - \sum_{k=1}^{n} f_k \tag{19}$$

The final F_k value for significance level is defined by the Equation (20).

$$F_k = \frac{SS_k / f_k}{SS_E / f_E}. \tag{20}$$

In general, the F value for significance level α (usually $\alpha = 0.05$ or $\alpha = 0.1$) should always be used as quantile points to judge if the factor k plays important role in effecting the results. Values of F_α could be obtained from the F distribution table. Comparing F_k obtained by calculating the F value to F_α from the F-table [32]: if $F_k > F_{0.05}$, the factor k is considered to have a highly significant effect on the test result; if $F_k > F_{0.1}$, the factor k is considered to have a significant influence on the test result; otherwise, it can be assumed that factor k has little effect on the results. The calculation results of ANOVA for all factors are shown in Table 7. From the common F-test table, $F_{0.05}(2, 2) = 19$ and $F_{0.1}(2, 2) = 9$ for the given degree of freedom shown in Table 7.

Table 7. Analysis of variance.

Source of Variance	$f(X)$ %			
	A	B	C	D
Sum of squared deviation SS_T	4.269	1.929	19.263	1.098
Degree of freedom f_T	2	2	2	2
Mean square error SS_E	1.1	1.1	1.1	1.1
Statistics value F_k	3.888	1.757	17.544	1
Significance	Insignificant	Insignificant	Significant	Insignificant

Level $\alpha = 0.1$, that is $F_\alpha (2, 2) = 9$, was used to judge the significance of the factor in this study. It can be concluded from Table 7 that the factor C has significant influence to pump performance, and the factors' order of importance is C, A, B, and D, which is consistent with the trend of the range analysis results. By comparing the t_i value, it can be determined that the superior levels of each factor are A3, B2, C3, and D3. According to the sensitivity analysis, the final optimized combination is A3B2C3D3 (scheme 2), and the specific parameters are impeller outlet diameter $D_2 = 46$ mm, blade outlet width $b_2 = 3.2$ mm, blade number $Z = 6$, blade outlet angle $\beta_2 = 35°$. The impeller used this optimization scheme compares with the original scheme, while numerical simulations were carried out using the same CFD process.

3.2.2. Analysis of Optimization Results

The comparison results of the optimized scheme 2 with the original scheme 1 is shown in Table 8. As seen from it, the heads of each assessment point are improved after optimization, in detail, the efficiency at design flowrate is increased by 3.66%, and the weighted average efficiency of the optimization scheme is increased by 3.29%.

Figure 11 shows the comparison of the external characteristic curves obtained from the CFD. It is clear that the head of the optimized pump is higher than the model pump, and the head curve is shifted to the direction of large flow. It can be seen from the η–Q curve that the flow rate at best efficiency point increased slightly. The efficiency improvement under large flow conditions is very obvious. Thus, the high efficiency region is effectively broadened.

Table 8. Comparison of the performances under specified conditions.

Operation Points	H/m		η/%		Weighted Average Efficiency %	
	Original	Optimal	Original	Optimal	Original	Optimal
$0.8Q_d$	7.6	7.9	50.1	52.17		
$1.0Q_d$	6.7	7.3	51.4	55.06	49.93	53.22
$1.25Q_d$	5.35	6.15	47.97	51.8		

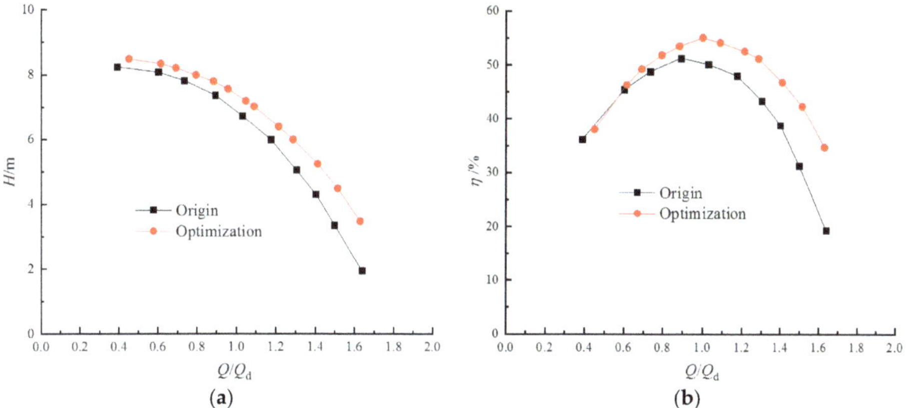

Figure 11. Comparison of external characteristic curves between origin and optimization scheme: (**a**) Head; (**b**) Efficiency.

Figure 12 shows the comparison of pressure distribution at mid-span of the impeller under three flowrates. Figure 12a–c represent the results of the original scheme and Figure 12d–f represent the results of the optimized scheme. As seen from them, the pressure inside the impeller passage increases from leading edge to trailing edge under all three different flowrates and appears largest near the impeller outlet. The pressure on the blade working surface is greater than the pressure on the back side of the corresponding position, and the low-pressure area appears on the suction side of the blade near the impeller inlet. As the flow rate increases, the area of the low-pressure area gradually expands. Under $1.25Q_d$, a reverse pressure gradient appears on the inlet part in the optimized impeller. The pressure distribution in the optimized pump impeller is more uniform than in the original one, and a relatively less obvious pressure gradient is present inside the impeller flow-path than the original one.

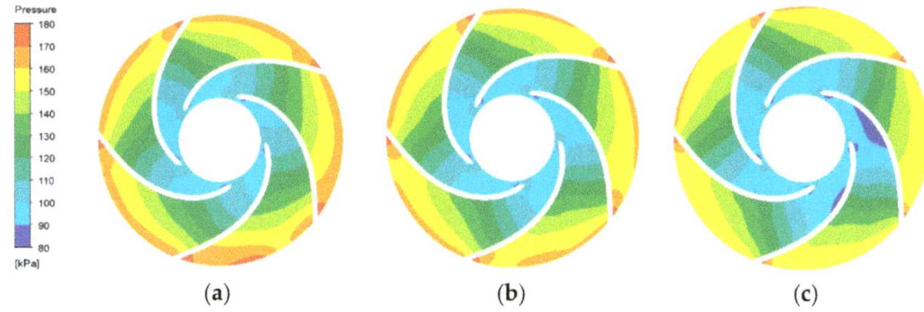

Figure 12. *Cont.*

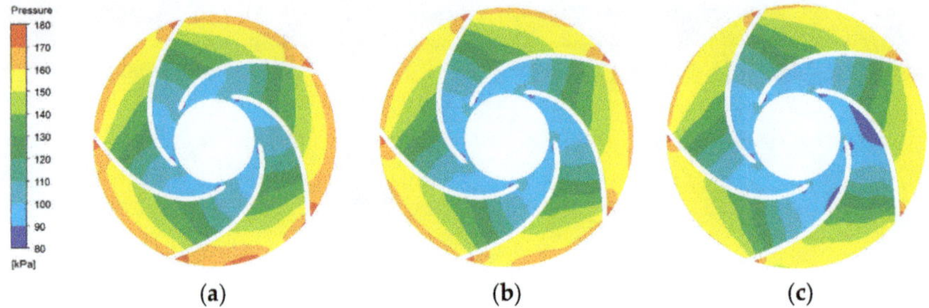

Figure 12. Comparison of pressure distribution in impeller for the two schemes: (**a**) origin, $0.8Q_d$; (**b**) origin, $1.0Q_d$; (**c**) origin, $1.25Q_d$; (**d**) optimization, $0.8Q_d$; (**e**) optimization, $1.0Q_d$; (**f**) optimization, $1.25Q_d$.

Figure 13 illustrates the distribution of the turbulence kinetic energy at mid-span of the impeller under three flowrates. Figure 13a–c represent the original scheme, Figure 13d–f represent the optimized scheme. It can be concluded that the maximum turbulence kinetic energy is prone to appear on the suction side of the blade. As the flow rate increases, the turbulence kinetic energy in the impeller gradually increases for both schemes. The turbulence kinetic energy of the optimized pump is smaller than the original one. Under $1.0Q_d$, the distribution of turbulence kinetic energy in the flow-path is the most uniform both before and after optimization. Under $1.25Q_d$, the kinetic energy distribution of the optimized pump is more even. Overall, the flow field has been improved after optimization.

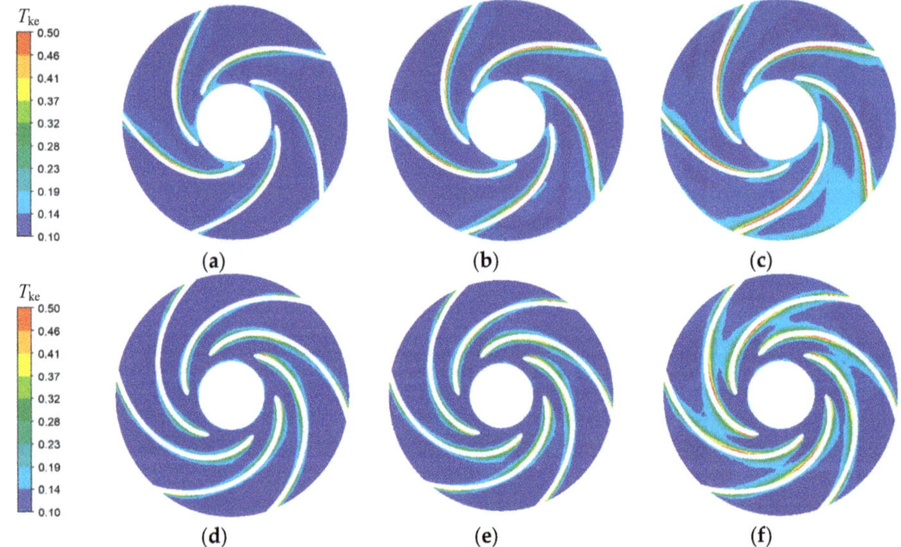

Figure 13. Comparison of turbulence kinetic energy in impeller for the two schemes: (**a**) origin, $0.8Q_d$; (**b**) origin, $1.0Q_d$; (**c**) origin, $1.25Q_d$; (**d**) optimization, $0.8Q_d$; (**e**) optimization, $1.0Q_d$; (**f**) optimization, $1.25Q_d$.

Figure 14 shows the comparison of the streamline in the volute at mid-span under three flowrates. Figure 14a–c represent the original scheme, Figure 14d–f represent the optimized scheme. It can be concluded that there is obvious flow separation in the volute outlet area accompanied by a large zone with low speed in the original scheme. Under the condition of $1.25Q_d$, a large low-speed vortex region appears in the volute outlet, and the flow state is very disordered, which generates large displacement to the mainstream. Under the $0.8Q_d$ and $1.0Q_d$ conditions, there are almost no obvious low-speed region and flow separation region in the outlet part of the optimized pump volute, and the flow velocity distribution is relatively uniform, indicating that the flow loss is small. Under $1.25Q_d$, flow separation phenomenon also occurs in the outlet part of the optimized pump. However, the size of the low-speed vortex region in the optimized pump is smaller than the original one, the velocity distribution in the spiral flow-path is more uniform, and the overall flow loss is greatly reduced. The optimized impeller improves the influence of rotor-stator interaction between the impeller and the volute tongue.

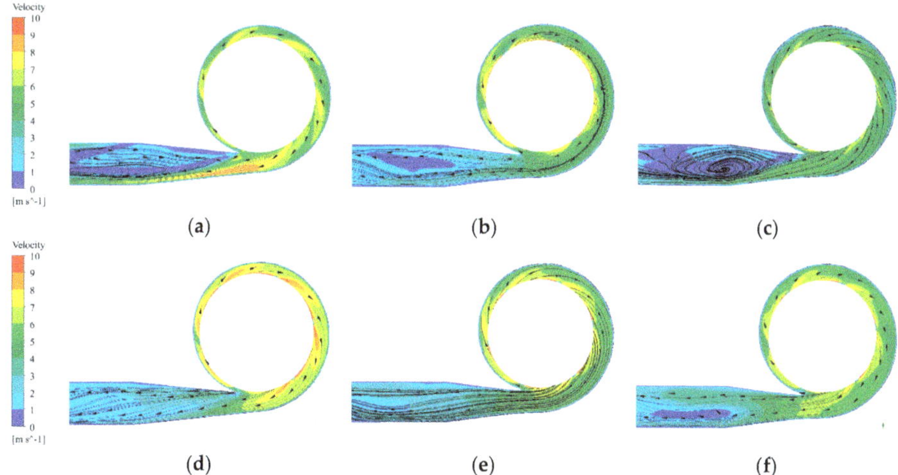

Figure 14. Comparison of velocity streamline in volute for the two schemes: (**a**) origin, $0.8Q_d$; (**b**) origin, $1.0Q_d$; (**c**) origin, $1.25Q_d$; (**d**) optimization, $0.8Q_d$; (**e**) optimization, $1.0Q_d$; (**f**) optimization, $1.25Q_d$.

3.3. Automatic Optimization with Intelligent Method

There is a complex multi-peak nonlinear relationship between the optimization objective and geometric parameters. Orthogonal design can only obtain a local optimal solution with fewer variables in discrete space. To further optimize the model pump in overall range quickly and efficiently, an intelligent algorithm program based on an automatic operation platform is needed [34]. Based on the analysis in the previous section, the selected optimization variables and the ranges are shown in Table 9, which includes the impeller diameter D_2, the width of impeller outlet b_2, inlet blade angle β_1, outlet blade angle β_2, leading edge tangential angle t_3, trailing edge tangential angle ϕ, and blade thickness δ. The variables did not include the parameter of Z at this step because this parameter plays the most important role in influencing the pump performance. Moreover, the F value of Z is much greater than the other three factors from Table 7, which means Z has significant influence. So, seven optimization variables with limited ranges in Table 9 were chosen to process the next optimization by intelligent method in order to reduce the scheme numbers.

Table 9. Ranges of optimization variables.

Parameters	Ranges
D_2(mm)	(45, 47)
b_2(mm)	(2.5, 3.5)
$\beta_1(°)$	(20, 30)
$\beta_2(°)$	(30, 40)
$t_3\ (°)$	(0, 25)
$\phi\ (°)$	(100, 120)
$\delta\ (\text{mm})$	(2, 6)

The genetic algorithm (GA), introduced by John Holland (1971), is a stochastic search technique based on natural selection and natural genetics mechanism to imitate living beings [35]. It has advantages in solving difficult optimization problems with high complexity and undesirable structure, which has been successfully used in fields of production planning and process optimization. The genetic process mainly includes selection, combination of crossover, and mutation. However, the robustness of the GA is not good enough to avoid premature convergence, which will lead to getting a local optimal solution instead of global one. In this study, an improved algorithm named the multi-island genetic algorithm (MIGA) is applied to solve this problem. The main feature of MIGA that distinguishes it from the traditional genetic algorithm is it divides each population of individuals into several subpopulations, namely 'islands'. All operations of the standard GA (selection, crossover, and mutation) are performed on each island separately. The used migration scheme is random ring, which means that the destination subpopulations are randomly chosen at every migration under the constraint that the migration is performed in sequential order between the subpopulations [36]. The structure and process of the MIGA optimization method is illustrated in Figure 15. MIGA selects individuals on each island to regularly migrate into other islands, and then continues with standard GA operations. The migration operation in the multi-island genetic algorithm keeps the diversity of knowledge, improves the chance of including the global optimal solution, and can suppress the occurrence of precocity. In that case, the MIGA is less likely to fall into local optimal than the GA method.

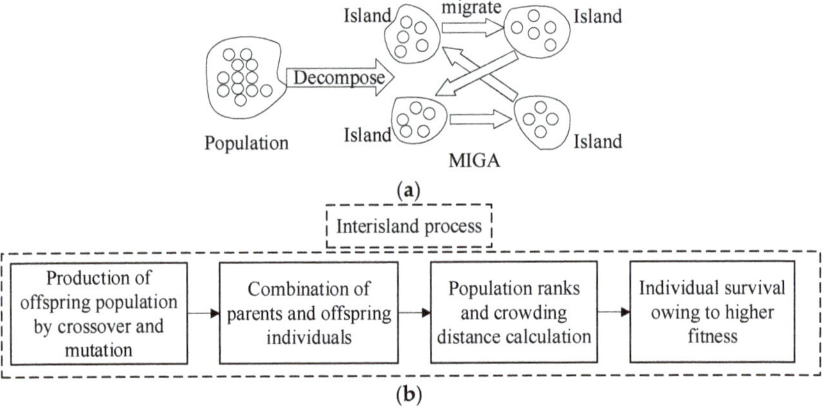

Figure 15. The structure and process of the multi-island genetic algorithm (MIGA) optimization method: (**a**) Structure, (**b**) Process.

Normally, parameters such as subpopulation size, number of islands, generations, crossover rate, migration rate, mutation rate have to be chosen appropriately. The main parameters of the selected MIGA are shown in Table 10. Total population size equal to the result of the subpopulation size multiplied by the number of islands, which is equal to 100 in this study (generally between 20 and 200). Because there are many optimization variables, the number of generations is set to 10, and then the necessary iteration for this optimization is 1000 steps. In order to improve the creating speed of the new individuals, the crossover rate is set to 0.9, and the mutation rate is set to 0.01 for maintaining the diversity of the population. The migration rate refers to the ratio of individual exchanges between each island, and the default value 0.01 in the software Isight is adopted. The interval of migration is also set to 5 by default.

Table 10. Optimization parameters adopted in MIGA.

Parameters	Numerical Value
Generations	10
Subpopulation Size	10
Number of Islands	10
Crossover rate	0.9
Migration rate	0.01
Interval of Migration	5
Mutation Rate	0.01

3.3.1. Optimization Results

Optimization based on the MIGA method was processed to obtain the global optimized scheme. Figure 16 shows the variation of the optimization target with the increase of iteration steps. The influence of the optimization variables on the efficiency of the automotive electronic water pump is complicated. The value of the optimization target fluctuates significantly with the increase of the iteration steps, and there are many schemes that do not satisfy the constraint conditions. As the iteration steps increase, the fluctuation value of the target becomes small. It means the multi-island genetic algorithm gradually locates the optimal solution region after a period of searching. The optimal solution (scheme 3) occurred at 798 iteration steps. All the calculations of the optimization works were done by using a normal desktop computer (Dell 7060MT) without running in parallel. The optimized hydraulic scheme was obtained within the limited 375 h.

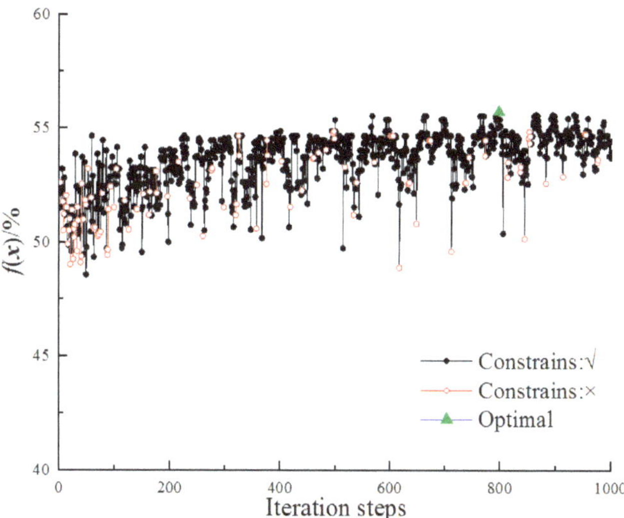

Figure 16. The iteration process of optimization.

Table 11 illustrates the comparison of geometric parameters between scheme 2 and scheme 3. As seen from the table, the outlet width of scheme 3 increases, the blade outlet angle and wrap angle both decrease slightly, and the impeller diameter becomes smaller. Figure 17 shows a comparison of the three-dimensional model of the two pump impellers.

Table 11. Comparison of geometric parameters between scheme 2 and scheme 3.

Parameters	Symbols	Unit	Scheme 2	Scheme 3
Outlet diameter	D_2	mm	46	45
Outlet width	b_2	mm	3.2	3.4
Blade number	Z	-	6	6
Blade outlet angle	β_1	°	22	26
Blade outlet angle	β_2	°	35	33
Leading edge tangential angle	t_3	°	0	8
Trailing edge tangential angle	ϕ	°	100	115
Blade thickness	δ	mm	2	2.4

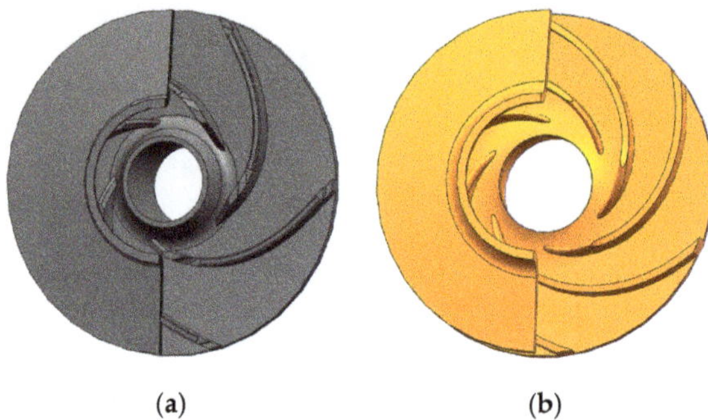

(a)   (b)

Figure 17. Comparison of impeller model: (**a**) scheme 1; (**b**) scheme 3.

Table 12 shows the head and efficiency of the optimized scheme 3 under three flowrates. It is clear that the head under the design flowrate has significantly improved compared with the original one. The weighted average efficiency under the three flowrates is 55.24%. The maximum efficiency of the pump is increased by 2%, the efficiency of the large flow condition is increased by 3.09%.

Table 12. Performance of optimized scheme (scheme 3).

	$0.8Q_d$	$1.0Q_d$	$1.25Q_d$
Head/m	8.57	8.15	6.86
Efficiency/%	52.98	57.06	54.89
$F(X)$/%		55.24	

3.3.2. Comparison of External Characteristic

Figure 18 shows the pump performance curve of the automotive electronic water pump obtained from the CFD. As seen from it, the head and efficiency of the optimized pump are obviously improved. The head curve of the optimized scheme 3 is always higher than the original head curve, and the efficiency don't improve much under small flow conditions. However, at the design flowrate, the efficiency is significantly improved, and also under the large flow rate. The efficiency is almost the same for the two schemes under small flow rate. It can be seen that the region with higher efficient working condition of the pump has been broadened for scheme 3, which means the area with high efficiency value located on the Q–η curve is further broadened after optimization.

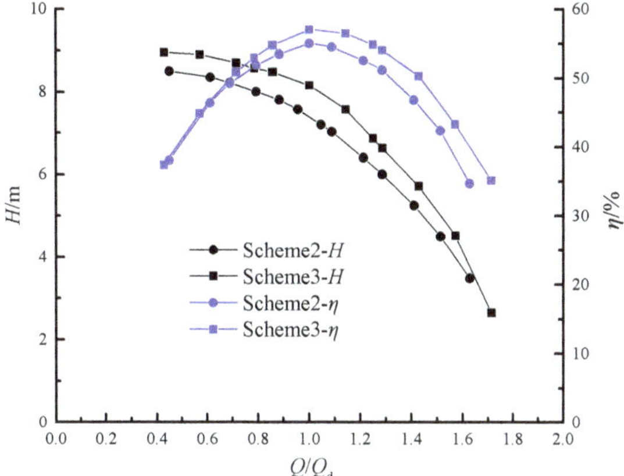

Figure 18. Comparison of external characteristic curves.

4. Optimization Verification

4.1. Comparison of Internal Flow Field

In this section, the flow characteristics of the cylindrical unfolding surface at different spans of the impeller are compared between the original scheme 1 and optimal scheme 3. The dimensionless distance of the impeller blade from hub to shroud is defined as:

$$\text{span} = \frac{r - r_h}{r_t - r_h} \times 100\%,\tag{21}$$

where r_t is the radius of shroud; r_h is the radius of hub; r is the radius of the cylindrical surface.

Figure 19 shows the comparison of velocity streamline at different spans under the design flowrate. It is obvious that a large area of flow separation occurs in the flow passages for scheme 1, large-scale separation vortices appear in the suction side, and the flow state becomes disordered, which seriously affects the performance of the pump. After optimization, the flow in the impeller passage shows a uniform state. The large-scale separation vortex disappears in the flow passages. Thus, the flow loss is reduced, which improves the performance of the pump.

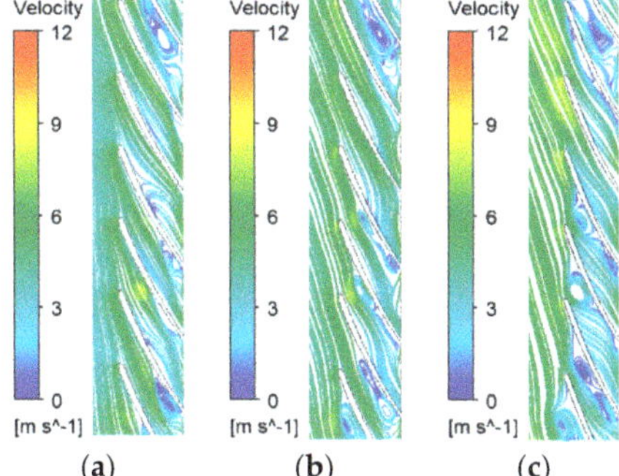

Figure 19. *Cont.*

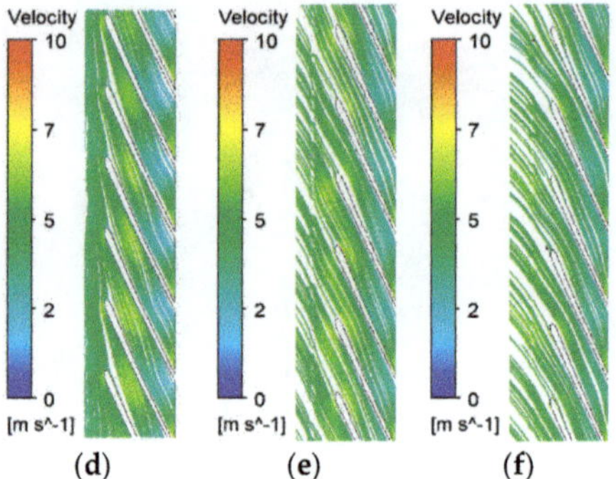

Figure 19. Comparison of streamline at different spans under $1.0Q_d$: (**a**) Origin scheme 1, span = 10%; (**b**) Origin scheme 1, span = 50%; (**c**) Origin scheme 1, span = 90%; (**d**) Optimal scheme 3, span = 10%; (**e**) Optimal scheme 3, span = 50%; (**f**) Optimal scheme 3, span = 90%.

Figure 20 shows the comparison of the pressure distribution at span = 50% under three flowrates. From the graph, the pressure distribution of the blade is lower at the wheel hub, the pressure at trailing edge location is higher. As the flow rate increases, the static pressure decreases. Under the same flow rate, the optimized pressure distribution diagram becomes more uniform than that before optimization. The pressure is gradually increased from the wheel hub to the rim, and the pressure at the front edge of the blade is effectively reduced.

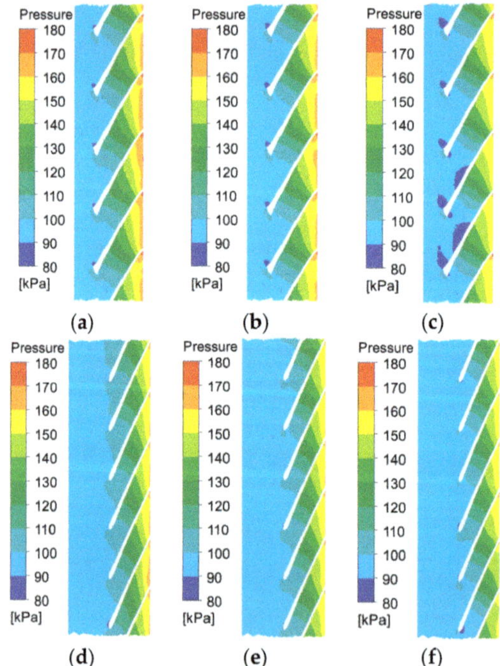

Figure 20. Pressure comparison at span = 90% under different working conditions: (**a**) Origin, $0.8Q_d$; (**b**) Origin, $1.0Q_d$; (**c**) Origin, $1.25Q_d$; (**d**) Optimal, $0.8Q_d$; (**e**) Optimal, $1.0Q_d$; (**f**) Optimal, $1.25Q_d$.

From the above analysis, the optimization scheme is superior to the original scheme in terms of internal flow.

4.2. Test Verification

In order to verify the reliability of the optimization results, the optimized impeller model (scheme 3) was 3D printed into a prototype for pump characteristic measurement. The test and the pump performance result curves compared with the original scheme are shown in Figure 21.

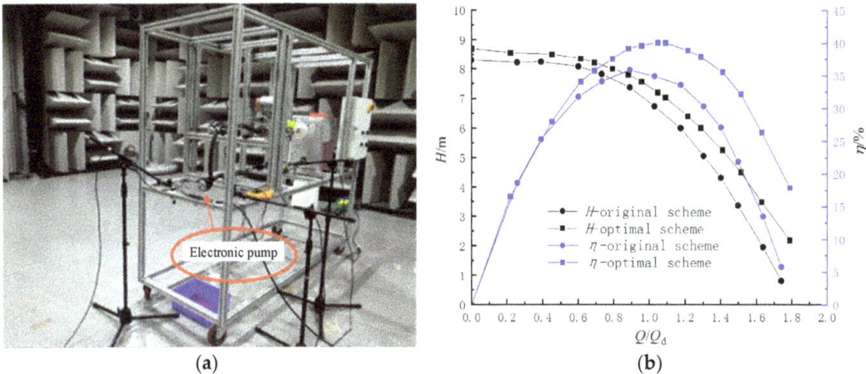

Figure 21. Comparison of the origin and optimal scheme. (**a**) Test rig; (**b**) Comparison pump performance results.

The efficiency obtained by the test of the automobile electronic water pump is the total efficiency of the pump system. As seen from Figure 21b, it is obvious that the head curve of the optimal scheme 3 is basically higher than the original scheme 1, and the downward trend is slower. The efficiency of the optimization is significantly improved by 4.3%. The efficiency curve of optimal scheme 3 declines slowly under a large flow rate, and the working range with high efficiency of the pump is broadened. The test results are basically consistent with the numerical simulation results, indicating that the optimization method in this study can be applied to the hydraulic optimization design of automotive electronic pumps.

5. Conclusions

A multi-point design process based on CFD and intelligent optimization method is proposed in this study to improve the energy transfer efficiency, taking the application of an automotive electronic pump as an example. Firstly, the three-dimensional CFD analysis of the prototype is carried out to understand the flow loss mechanism inside the pump and establish the numerical prediction model of pump performance. Secondly, an automatic optimization platform including fluid domain modeling, meshing, solving, post-processing, and design of experiment (DOE) is built based on the three-dimensional parametric design method. Then, orthogonal experimental design and the multi-island genetic algorithm (MIGA) were utilized to drive the platform for improving the efficiency of the pump at three flowrates. Finally, the optimal impeller geometries were obtained and manufactured into a prototype for verification. The conclusions are as follows.

(1) After orthogonal optimization, the heads of each working point are significantly improved. The weighted average efficiency of the optimization scheme 2 is improved by 3.29%. The number of impeller blades has the most important effects on pump performance improvement. After the intelligent optimization, the high-efficiency region of the automotive electronic water pump is further widened. The efficiency weighted average recorded 55.24% under three working conditions. The optimal efficiency at design flowrate is increased by 4.3% after optimization from the experiment test.

(2) There is almost no obvious low-speed region under the optimal conditions and small flow conditions after orthogonal optimization. The flow loss is greatly reduced in the impeller and volute flow-path, the number of separation vortices and their area under a large flowrate are

smaller than the original model pump. The pressure distribution in the impeller flow-path of the optimized pump is more uniform, and the pressure gradient becomes smaller inside the impeller flow-path. the overall turbulence energy is significantly reduced after optimization. All of the above is the root cause of the efficiency improvement.

(3) The automatic optimization platform built in this study combined with intelligent MIGA algorithm could obtain the global optimization scheme among the selected parameter range. The hydraulic performance of automotive electronic water pumps under three multiple operating conditions have significant improvement, realizing the integration of "design-simulation-optimization" in limited time, which means this method has great potential in the application of fluid machinery design.

Author Contributions: Methodology, Q.S. and J.Y.; simulation, G.S.; analyzed the data, R.L. and C.S.; writing—review and editing, Q.S.; funding acquisition and validation, S.X. All authors have read and agreed to the published version of the manuscript.

Funding: This research was funded by National Key Research and Development Program of China (2018YFB0606103), National Natural Foundation of China (51976079, 51779107) and China Postdoctoral Science Foundation (2019M661745).

Acknowledgments: The authors gratefully acknowledge the financial support by National Key Research and Development Program of China (2018YFB0606103), National Natural Foundation of China (51976079, 51779107) and China Postdoctoral Science Foundation (2019M661745).

Conflicts of Interest: The authors declare no conflict of interest.

References

1. Shankar, V.K.; Umashankar, S.; Paramasivam, S.; Hanigovszki, N. A comprehensive review on energy efficiency enhancement initiatives in centrifugal pumping system. *Appl. Energy* **2016**, *181*, 495–513. [CrossRef]
2. Vlad, H.; Alin, I.B.; Sébastien, L.; Cécile, M.A. A Dynamic Approach for Faster Performance Measurements on Hydraulic Turbomachinery Model Testing. *Appl. Sci.* **2018**, *8*, 1426. [CrossRef]
3. Cortona, E.; Onder, C. Engine thermal management with electric cooling pump. *SAE Tech. Pap. Ser.* **2000**, *1*, 0965. [CrossRef]
4. Hoon, C.; Dohoy, J.; Zoran, S.F.; Dennis, N.A.; John, V.; Walter, B. Application of controllable electric coolant pump for fuel economy and cooling performance improvement. *J. Eng. Gas Turbines Power* **2007**, *129*, 43–50.
5. Wang, X.; Liang, X.; Hao, Z.; Chen, R. Comparison of electrical and mechanical water pump performance in internal combustion engine. *Int. J. Veh. Syst. Model. Test.* **2015**, *10*, 205–223. [CrossRef]
6. Si, Q.R.; Yuan, S.Q.; Yuan, J.P.; Wang, C.; Lu, W.G. Multi-objective optimization of low-specific-speed multistage pumps by using matrix analysis and CFD method. *J. Appl. Math.* **2013**, *10*, 261–274.
7. Kim, K.Y.; Samad, A.; Benini, E. *Design Optimization of Fluid Machinery-Applying Computational Fluid Dynamics and Numerical Optimization*; John Wiley & Sons, Inc.: Hoboken, NJ, USA, 2019.
8. Wang, C.; He, X.; Shi, W.; Wang, X.; Wang, X.; Qiu, N. Numerical study on pressure fluctuation of a multistage centrifugal pump based on whole flow field. *AIP Adv.* **2019**, *9*, 035118. [CrossRef]
9. Kim, J.H.; Lee, H.C.; Kim, J.H.; Choi, Y.S.; Yoon, J.Y.; Yoo, I.S.; Choi, W.C. Improvement of hydrodynamic performance of a multiphase pump using design of experiment techniques. *J. Fluids Eng. Trans. ASME* **2015**, *137*, 081301. [CrossRef]
10. Wu, D.; Yan, P.; Chen, X. Effect of trailing-edge modification of a mixed-flow pump. *J. Fluids Eng. Trans. ASME* **2015**, *137*, 1–9. [CrossRef]
11. Zhou, L.; Shi, W.; Lu, W. Numerical investigations and performance experiments of a deep-well centrifugal pump with different diffusers. *J. Fluids Eng. Trans. ASME* **2012**, *134*, 071102. [CrossRef]
12. Osman, M.K.; Wang, W.; Yuan, J.; Zhao, J.; Wang, Y.; Liu, J. Flow loss analysis of a two-stage axially split centrifugal pump with double inlet under different channel designs. *Proc. Inst. Mech. Eng. C J. Mech. Eng. Sci.* **2019**, *233*, 5316–5328. [CrossRef]
13. Wang, C.; Shi, W.; Wang, X. Optimal design of multistage centrifugal pump based on the combined energy loss model and computational fluid dynamics. *Appl. Energy* **2017**, *187*, 10–26. [CrossRef]
14. Liu, M.; Tan, L.; Cao, S. Design method of controllable blade angle and orthogonal optimization of pressure rise for a multiphase pump. *Energies* **2018**, *11*, 1048. [CrossRef]

15. Li, X.; Liu, Z.; Lin, Y. Multipoint and multiobjective optimization of a centrifugal compressor impeller based on genetic algorithm. *Math. Probl. Eng.* **2017**, *2017*, 1–18. [CrossRef]
16. Luo, W.; Lyu, W. An application of multidisciplinary design optimization to the hydrodynamic performances of underwater robots. *Ocean Eng.* **2015**, *104*, 686–697. [CrossRef]
17. Aziz, M.; Tayarani, N.M.H. An adaptive memetic particle swarm optimization algorithm for finding large-scale latin hypercube designs. *Eng. Appl. Artif. Intell.* **2014**, *36*, 222–237. [CrossRef]
18. Michalewicz, Z.; Krawczyk, J.B.; Kazemi, M. Genetic Algorithms and Optimal Control Problem. In Proceedings of the 29th IEEE Conference on Decision and Control, San Francisco, CA, USA, 5–7 December 1990.
19. Pei, J.; Wang, W.; Yuan, S. Multi-point optimization on meridional shape of a centrifugal pump impeller for performance improvement. *J. Mech. Sci. Technol.* **2016**, *30*, 4949–4960. [CrossRef]
20. Pei, J.; Wang, W.; Osman, M.K.; Gan, X. Multiparameter optimization for the nonlinear performance improvement of centrifugal pumps using a multilayer neural network. *J. Mech. Sci. Technol.* **2019**, *33*, 2681–2691. [CrossRef]
21. Yuan, S.Q.; Wang, W.J.; Pei, J.; Zhang, J.F.; Mao, J.Y. Multi-objective optimization of low-specific-speed centrifugal pump. *Trans. Chin. Soc. Agric. Eng.* **2015**, *31*, 46–52.
22. Zhang, Y.; Hu, S.; Wu, J.; Zhang, Y.; Chen, L. Multi-objective optimization of double suction centrifugal pump using Kriging metamodels. *Adv. Eng. Softw.* **2014**, *74*, 16–26. [CrossRef]
23. Wang, C.; Ye, J.; Zeng, C.; Xia, Y.; Luo, B. Multi-objective optimum design of high specific speed mixed-flow pump based on NSGA-II genetic algorithm. *Trans. Chin. Soc. Agric. Eng.* **2015**, *31*, 100–106.
24. Zhang, J.; Zhu, H.; Li, Y. Shape Optimization of Helico-Axial Multiphase Pump Impeller Based on Genetic Algorithm. In Proceedings of the 2009 Fifth International Conference on Natural Computation, Tianjin, China, 14–16 August 2009.
25. Yang, X.S. Firefly algorithms for multimodal optimization. In *Stochastic Algorithms: Foundations and Applications, SAGA 2009, Lecture Notes in Computer Sciences*; Springer: Berlin/Heidelberg, Germany, 2009; Volume 5792, pp. 169–178.
26. Shao, Y.; Ou, H.; Guo, P.; Yang, H. Shape optimization of preform tools in forging of aerofoil using a metamodel-assisted multi-island genetic algorithm. *J. Chin. Inst. Eng.* **2019**, *42*, 297–308. [CrossRef]
27. Wang, Q.X.; Wang, H.; Qi, Z.Q. An application of nonlinear fuzzy analytic hierarchy process in safety evaluation of coal mine. *Saf. Sci.* **2016**, *86*, 78–87. [CrossRef]
28. Ruiz-Padillo, A.; Torija, A.J.; Ramos-Ridao, A.F.; Ruiz, D.P. Application of the fuzzy analytic hierarchy process in multi-criteria decision in noise action plans: Prioritizing road stretches. *Environ. Model. Softw.* **2016**, *81*, 45–55. [CrossRef]
29. Chaudhary, P.; Chhetri, S.K.; Joshi, K.M. Application of an analytic hierarchy process (AHP) in the GIS interface for suitable fire site selection: A case study from kathmandu metropolitan city, Nepal. *Socio Econ. Plan. Sci.* **2016**, *53*, 60–71. [CrossRef]
30. Zhang, J.; Fu, S. An efficient approach for quantifying parameter uncertainty in the SST turbulence model. *Comput. Fluids* **2019**, *181*, 173–187. [CrossRef]
31. Wang, W.; Mo, R.; Zhang, Y. Multi-objective aerodynamic optimization design method of compressor rotor based on Isight. *Procedia Eng.* **2011**, *15*, 3699–3703. [CrossRef]
32. Kim, J.H.; Choi, J.H.; Husain, A.; Kim, K.Y. Performance enhancement of axial fan blade through multi-objective optimization techniques. *J. Mech. Sci. Technol.* **2010**, *24*, 2059–2066. [CrossRef]
33. Balaram, N.A.; Chennakeshava, R.A. Optimization of tensile strength in TIG welding using TAGUCHI method and analysis of variance (ANOVA). *Therm. Sci. Eng. Prog.* **2018**, *8*, 327–339. [CrossRef]
34. Hu, X.; Chen, X.; Zhao, Y.; Zhao, Y. Optimization design of satellite separation systems based on Multi-Island Genetic Algorithm. *Adv. Space Res.* **2014**, *53*, 870–876. [CrossRef]
35. Sinha, A.; Malo, P.; Deb, K. Evolutionary algorithm for bilevel optimization using approximations of the lower level optimal solution mapping. *Eur. J. Oper. Res.* **2016**, *257*, 395–411. [CrossRef]
36. Wang, W.J.; Yuan, S.Q.; Pei, J.; Zhang, J.F. Optimization of the diffuser in a centrifugal pump by combining response surface method with multi-island genetic algorithm. *Proc. IMechE Part. E J. Process. Mech. Eng.* **2015**, *231*, 191–201. [CrossRef]

Article

Integrated Optimum Layout of Conformal Cooling Channels and Optimal Injection Molding Process Parameters for Optical Lenses

Chen-Yuan Chung

Department of Mechanical Engineering, National Central University, Taoyuan City 32001, Taiwan; cxc474@case.edu or cychung@ncu.edu.tw; Tel.: +886-34267331; Fax: +886-34254501

Received: 20 August 2019; Accepted: 12 October 2019; Published: 15 October 2019

Abstract: Plastic lenses are light and can be mass-produced. Large-diameter aspheric plastic lenses play a substantial role in the optical industry. Injection molding is a popular technology for plastic optical manufacturing because it can achieve a high production rate. Highly efficient cooling channels are required for obtaining a uniform temperature distribution in mold cavities. With the recent advent of laser additive manufacturing, highly efficient three-dimensional spiral channels can be realized for conformal cooling technique. However, the design of conformal cooling channels is very complex and requires optimization analyses. In this study, finite element analysis is combined with a gradient-based algorithm and robust genetic algorithm to determine the optimum layout of cooling channels. According to the simulation results, the use of conformal cooling channels can reduce the surface temperature difference of the melt, ejection time, and warpage. Moreover, the optimal process parameters (such as melt temperature, mold temperature, filling time, and packing time) obtained from the design of experiments improved the fringe pattern and eliminated the local variation of birefringence. Thus, this study indicates how the optical properties of plastic lenses can be improved. The major contribution of present proposed methods can be applied to a mold core containing the conformal cooling channels by metal additive manufacturing.

Keywords: gradient-based algorithm; robust genetic algorithm; warpage; design of experiments; fringe pattern; birefringence

---

1. Introduction

The demand for plastic optical lenses has been increasing in the industry. The precision requirements for high-tech optical products have become stringent, which has led to the growth of global markets for high-precision optical articles [1,2]. Although the optical properties of glass, such as the refractive index and dispersion, are quite stable, the glass fabrication process is very complicated and difficult. Moreover, plastic products are lightweight, colorable, robust, and low cost. They can be manufactured using a one-step process regardless of their geometric complexity. Thus, plastic products have become crucial in contemporary industrial development.

Despite the aforementioned advantages, plastic optical lenses may encounter volumetric shrinkage, which leads to the formation of thermally induced residual stress during the cooling process of injection molding. This residual stress slightly results in local variations in the birefringence, which affects the image quality [3]. Achieving a uniform temperature distribution for removing the residual stress is difficult in conventional cooling channels. Uneven shrinkage occurs if conventional cooling channels are used during the cooling process [4]. For designing conformal cooling channels, the geometric shape of conventional cooling channels can be appropriately adjusted through injection molding simulation. This helps to improve the defects caused by conventional cooling channels. The use of conformal cooling channels facilitates an even distribution of the surface temperature of the mold cavities, thereby reducing the thermally induced residual stress formed during the cooling process [5], effectively shortening the cooling time, and improving the cooling efficiency [6].

Researchers have devoted considerable attention to hybrid manufacturing processes, which combine metallic powder-based laser additive processes and subtractive machining processes. These hybrid manufacturing processes are considered the most promising technology for fabricating conformal cooling channels [7]. With metal additive manufacturing technology, complex conformal cooling channels can be fabricated to closely fit the shape of the mold cavity and core. Thus, uniform cooling can be achieved for products even in narrow regions or areas that may easily accumulate heat. Consequently, the quality of the products improves, and the cycle time decreases. This technology can be applied to products that vary in thickness. It allows the products to achieve uniform heat dissipation with a high cooling efficiency [8].

In recent years, optimization algorithms have been used to design cooling channels for plastic injection molds. Qiao [9] combined the advantages of the Davidon–Fletcher–Powell (DFP) method and the simulated annealing (SA) algorithm to optimize a cooling system layout. First, the DFP method was used to find the local optimum layout of the cooling channel. Then, the SA algorithm was adopted to determine the global optimum layout of the cooling channel, which allowed the surfaces of the mold cavities to possess a uniform temperature distribution. Park and Dang [10] used design of experiments (DOE) and response surface methodology for designing an array of baffles in cooling channels. They established a mathematical model for obtaining the optimal configuration of cooling channels with an array of baffles. This technique can effectively improve the heat removal performance and is applicable to large-sized molds and molds with complex cavity shapes. Dang and Park [11] also proposed an optimization method for the design of U-shaped milled groove cooling channels. This method aimed to achieve temperature uniformity for the mold and utilized computer-aided engineering (CAE) for design modification. The mold of a car fender was selected to analyze its cooling channel design and verify the theoretical calculation. The quality levels of the products were compared before and after the optimization. After the optimization of the cooling channels, the warpage decreased, and the temperature uniformity increased.

Optimization strategies involving the use of various algorithms for designing the shape and layout of cooling channels have been frequently discussed in the literature [12–14]. Moreover, many studies have used DOE to set optimal process parameters for injection molding [15,16]. However, these two issues have rarely been integrated in the literature. This study combined finite element analysis with optimization algorithms to analyze the temperature field during the cooling stage. Subsequently, the entire injection molding process was conducted using DOE to obtain the best process parameters. The aim of this study was to uniformly cool melt within a cavity. An optimization was conducted to design conformal cooling channels, which alleviated the thermally induced residual stress formed during the manufacturing of optical lenses and solved the uneven shrinkage problem. Such optimization can also effectively shorten the cooling time and enhance the cooling efficiency during the manufacturing process, which can improve the image quality of plastic optical lenses.

2. Methods

2.1. Materials

The lens employed in this study was a plastic optical lens commonly used in projectors. This projector lens was a large-diameter aspheric lens [17] with a diameter of 46 mm (Figure 1). The lens was composed of cyclo-olefin polymer (COP; Zeonex 480R) from the Zeon Corporation (Tokyo, Japan). This material had low water absorption, high optical transmittance, and low birefringence [18].

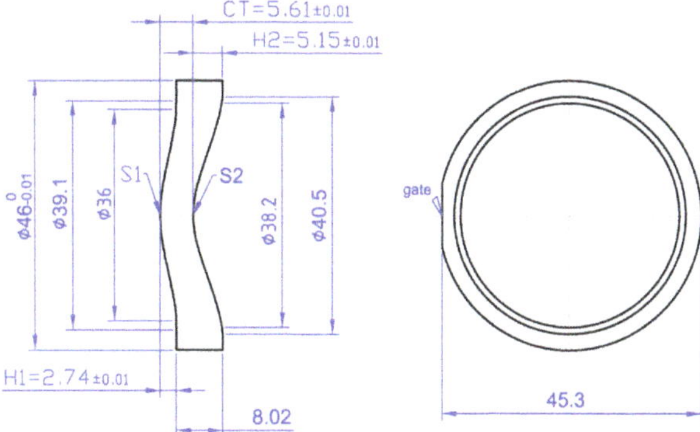

Figure 1. Specifications of the large-diameter aspheric plastic lens provided by Glory Science Company Limited.

2.2. Design of the Runner and Gate System

Figure 2 displays the shape and dimensions of the cold runner and gate system. The volume of the runner and gate system was 8.984 cm^3. A uniform melt flow front was injected into the mold cavity by using the wide cross-sectional inlet of the fan gate. This enabled a molded product with large width to be filled quickly. The warpage and size stability of wide molded products are major concerns [4]. Although the fan gate necessitated manual trimming, it reduced the formation of flow-induced residual stress when the melt polymer passed through the gate into the mold cavity. Therefore, the fan gate was adopted in this study. Residual stress may lead to poor optical properties, such as uneven distribution of the fringe pattern and local variation in the birefringence.

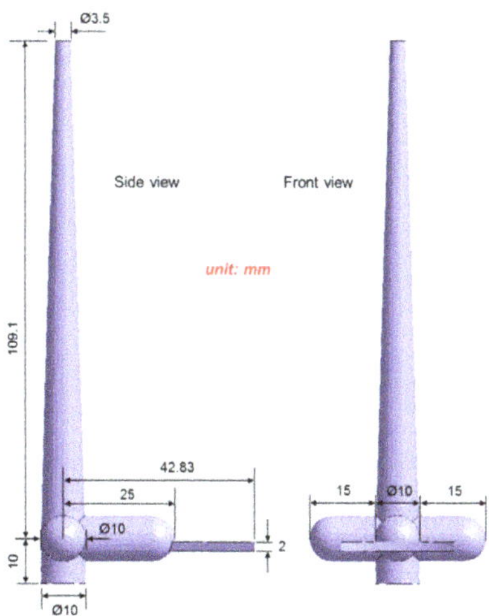

Figure 2. Shape and dimensions of the runner–gate system.

2.3. Mold Design for Conventional and Planar Conformal Cooling Channels

The core and cavity plates used in this study were made of NAK80, which is prehardened steel. The dimensions of the mold were 150 mm (L) × 150 mm (W) × 205 mm (H). A single-cavity mold containing both conventional cooling channels and planar conformal cooling channels was adopted (Figure 3) because the lens was large. The total number of elements in the cavity was 222,720. The conventional cooling channel diameter was suggested in the design [4] to be 10 mm. The layout of the planar conformal cooling channel adhered to the channel design guidelines proposed in the literature [19] for laser additive manufacturing of metals. These rules were used to determine the appropriate distance between channels or between the channel and mold cavity surface according to the cooling channel diameter (Figure 4). Following these rules was the way to ensure that the mold cavity and core have enough mechanical strength. To alleviate the thermally induced warpage [20], the parameters of the planar conformal cooling channel were set as follows: b = 4 mm, a = 8 mm, and c = 8 mm, as displayed in Figure 5.

Figure 3. Internal arrangement of the mold used in this study. Deep blue, bright blue, green, and orange components represent the conventional cooling channels, planar conformal cooling channels, lens, and runner–gate, respectively.

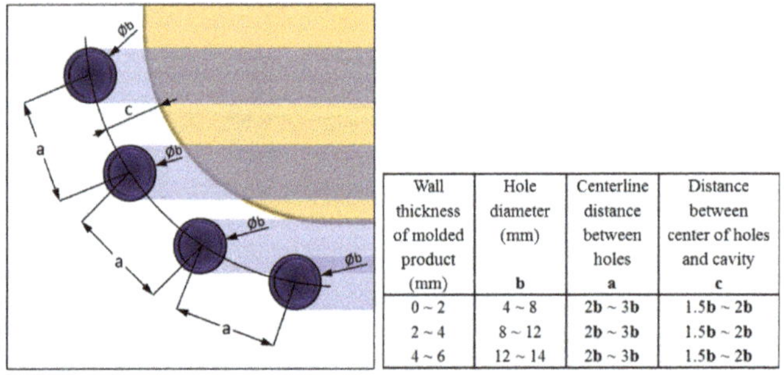

Wall thickness of molded product (mm)	Hole diameter (mm) b	Centerline distance between holes a	Distance between center of holes and cavity c
0 ~ 2	4 ~ 8	2b ~ 3b	1.5b ~ 2b
2 ~ 4	8 ~ 12	2b ~ 3b	1.5b ~ 2b
4 ~ 6	12 ~ 14	2b ~ 3b	1.5b ~ 2b

Figure 4. Design rules for the three-dimensional conformal cooling channel [19].

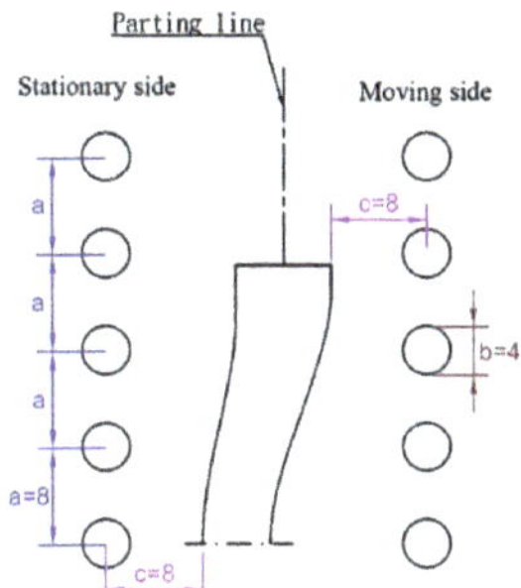

Figure 5. Schematic of the planar conformal cooling channel.

2.4. Simulation of the Filling and Packing Stages

In this study, the CAE mode of Moldex3D was used to simulate injection molding because no empirical data were available for the machine settings. The maximum injection pressure and maximum packing pressure were set as 250 MPa. The initial melt temperature was set as 270 °C, and the initial mold temperature was set as 100 °C according to the process parameters [21,22] from the Moldex3D (CoreTech System Corporation, Taiwan) databank. The filling time was set as 0.79 s, and the packing time was set as 7 s. The filling flow rate was set in six steps (Figure 6a). The first step involved the process of filling the runner. The subsequent five steps varied according to the variations in the cross-sectional area of the plastic lens. The packing pressure was set in three steps (Figure 6b). The packing pressure of the first step was set as 85% of the filling pressure at the end of filling. In the following two steps, the packing pressure was decreased to release stress.

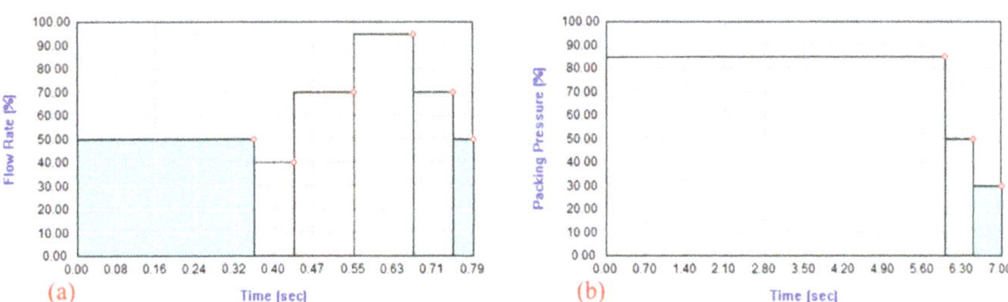

Figure 6. (**a**) Multistep setting of the filling flow rate profile and (**b**) multistep setting of the packing pressure profile.

2.5. Geometric Optimization of the Conformal Cooling Channels

The average mold temperature was 89.5 °C and the average melt temperature was 215.6 °C at the completion of the Moldex3D simulation of the filling and packing stages. These temperatures were input into COMSOL Multiphysics (Version 5.2, COMSOL Inc., Burlington, MA, USA) and served as the initial conditions of the cooling stage. According to the default settings in Moldex3D, the suggested channel temperature was 100 °C, and the suggested cooling time was 18.6 s. However, the cooling time in the COMSOL software was set as 20 s, and neither the runner nor the mold base was included in the simulation. The simulation only focused on the heat transfer between the cooling channel and lens. The cooling channel was assumed to have a turbulent flow. By using the Reynolds number formula, the volumetric flow rate of the conventional cooling channel with a diameter of 10 mm was derived as 28.81 cm^3/sec and that of the planar conformal cooling channel with a diameter of 4 mm was derived as 11.53 cm^3/sec. Moreover, COMSOL was adopted to couple the non-isothermal pipe flow interface with the heat transfer in solids interface [23] for simulating the temperature distribution of the melt in the mold cavity during the cooling process. The solidification of the polymer melt flow near the cold cavity wall was not considered. Furthermore, two approaches in SmartDO (FEA-Opt Technology Inc., Miaoli County; Taiwan), namely the gradient-based algorithm (GBA) [24] and robust genetic algorithm (RGA) [25,26], were separately integrated with COMSOL to optimally design the layout of the conformal cooling channels. The conventional cooling channels were not involved in optimization. The integration framework is displayed in Figure 7. The script loop is available in the online supplementary data. The pros and cons of the two optimization algorithms were compared according to the temperature distributions on the lens surfaces.

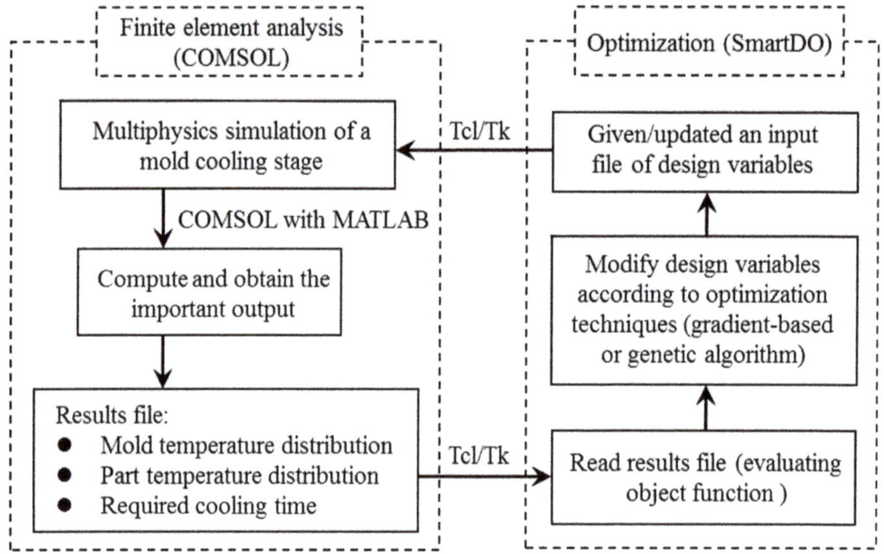

Figure 7. Flow chart of finite element analysis integrated with the optimization algorithms, gradient-based algorithm (GBA) and robust genetic algorithm (RGA), for the design of conformal cooling channels.

The value of the distance between the conformal cooling channel and mold cavity surface must satisfy the rules suggested in Figure 4 so that the design variables (DVs) have a reasonable range, as displayed in Figure 8. Moreover, at the suggested cooling time of 18.6 s, COMSOL indicated that the maximal and minimal surface temperatures of the mold cavity for the planar conformal cooling channels were $T^0_{sur,max}$ and $T^0_{sur,min}$, respectively. After each iteration of the optimization, the maximal and minimal surface temperatures of the mold cavity for the modified conformal cooling channels were $T_{sur,max}$ and $T_{sur,min}$, respectively. To ensure that the temperature of mold cavity surface was evenly distributed (i.e., the value of $T_{sur,max} - T_{sur,min}$ was small), the objective function was set as following Equation (1):

$$[1 - (T_{sur,min} / T_{sur,max})] \times 10,000 \tag{1}$$

The scaling factor of 10,000 in Equation (1) increases the accuracy of the calculation. To ensure that the overall temperature of the mold cavity surface after optimization was lower than that of the initial state, the following constraints were set:

$$T_{sur,max} - T^0_{sur,max} < 0 \tag{2}$$

$$T_{sur,min} - T^0_{sur,min} < 0 \tag{3}$$

After setting the DVs, objective function, and constraints, the planar conformal cooling channels could be designed and modified as three-dimensional channels by using the optimization algorithm. This modification improved the cooling efficiency and allowed the temperature of the melt in the mold cavity to be evenly distributed.

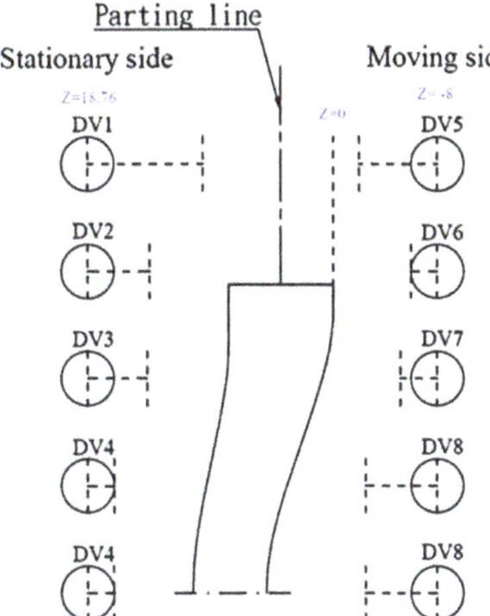

Design Variable	Initial Value	Upper Bound	Lower Bound
DV1	18.76	18.76	10.01
DV2	18.76	18.76	14.02
DV3	18.76	18.76	14.192
DV4	18.76	18.76	16.76
DV5	-8	-1.99	-8
DV6	-8	-6	-8
DV7	-8	-5.176	-8
DV8	-8	-2.467	-8

Figure 8. Lower and upper bounds on the design variables (DVs) according to the distances suggested in Figure 4 between the cooling channel and mold cavity surface, where the initial value represents the geometric layout of the planar conformal cooling channels.

2.6. Molding Process Optimization for the Optimized Conformal Cooling Channels

After obtaining the optimized DVs for the conformal cooling channels by using the GBA and RGA algorithms, the geometric layouts of these two types of cooling channels were converted into solid mesh models by using the Rhinoceros 3D software and were then imported into Moldex3D. The process conditions for the filling and packing stages were the same as those mentioned in Section 2.4. The cooling time, ejection temperature, and mold-open time were set as 18.6 s, 139 °C, and 5 s, respectively. After the entire injection molding process had been simulated, the three types of conformal cooling channels (the planar, GBA- and RGA-optimized channels) were compared with regards to warpage and temperature distribution.

Finally, the geometric layout with the highest cooling efficiency was selected for DOE to determine the optimal combination of process parameters and key parameters impacting the manufacturing process. Studies [21,22] have indicated that lenses made of Zeonex 480R material have high flow-induced birefringence and low thermally induced birefringence. This study focused on the shear stress related to flow and the flow-induced residual stress. Residual stress leads to defective optical properties [16]. Therefore, shear stresses at the end of the filling stage, shear stresses at the end of the packing stage, and the total warpage were set as the quality factors, and each quality characteristic was based on the smaller-the-better approach. The traditional trial and error method for predicting and controlling injection molding conditions is inefficient and costly because of the complexity of interactions between multiple manufacturing process parameters. Therefore, the DOE module provided by Moldex3D is more suitable than the trial and error method for evaluating the ideal molding conditions [15]. Moreover, the melt temperature, mold temperature, filling time, and packing time were selected as the control factors in the DOE. It was assumed that each control factor contained five levels of variation (Table 1). A Taguchi's orthogonal array $L_{25}(5^4)$ [27] was adopted for the DOE. Subsequently, statistical analysis was used to determine the optimum combination of levels for the control factors.

Table 1. Four control factors and five levels used for the design of experiments (DOE) in Moldex3D.

	Control factors	Level 1	Level 2	Level 3	Level 4	Level 5
A	Melt temperature (°C)	240	255	270	285	300
B	Mold temperature (°C)	80.00	89.75	99.50	109.25	119.00
C	Filling time (sec)	0.69	0.79	0.89	0.99	1.09
D	Packing time (sec)	6	7	8	9	10

3. Results and Discussion

In this research, COMSOL software simulated the cooling process of injection molding. Moreover, SmartDO optimization software was used to design the conformal cooling channels for improving the temperature distribution of the melt and the cooling efficiency. Furthermore, Moldex3D software was used to simulate a complete injection molding cycle at the same manufacturing conditions. The temperature difference between the inlet and outlet, average surface temperature of the lens, and warpage deformation of the lens were investigated for the three types of conformal cooling channels (the planar, GBA- and RGA-optimized channels). Finally, the optimum layout with the highest cooling efficiency was selected to optimize the molding conditions by using a DOE module. This optimization improved the birefringence and the fringe pattern of the lens.

3.1. Optimum Layout of Conformal Cooling Channels for the Cooling Stage

According to the Moldex3D simulation, the average mold temperature was 89.5 °C and the average melt temperature was 215.6 °C at the end of the packing stage. These temperatures were input into COMSOL and served as the initial conditions of the cooling stage. By simulating the cooling process for the planar conformal cooling channels, the variations in the maximum and minimum surface temperature of the melt in the mold cavity were determined (Figure 9a). According to the cooling time suggested by Moldex3D, the lens was ejected at 18.6 s. Figure 9b displays the surface temperature distribution for the lens, with a maximum temperature of 359.713 K, a minimum temperature of 358.601 K, and a temperature difference of 1.112 K.

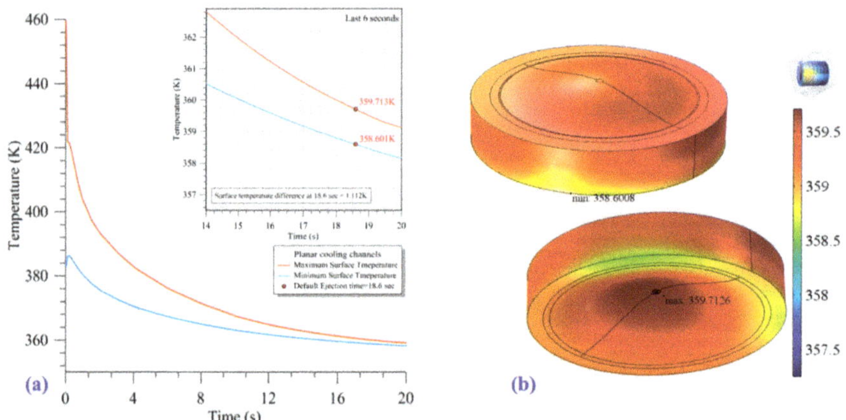

Figure 9. (**a**) Surface temperature of the melt during the cooling stage and (**b**) surface temperature of the melt at the default ejection time of 18.6 s when using the planar conformal cooling channels.

The GBA and RGA algorithms of the SmartDO software were employed for evenly reducing the surface temperature of the melt in the mold cavity. The objective function of the two algorithms converged to a minimum value in the optimization procedure (Figure 10). The objective function value of the GBA was 19.78, whereas that of the RGA was 16.54. The optimized design parameters obtained using these two algorithms are displayed in Figure 11. By using these design parameters (Table 2), two conformal cooling channels with different geometric layouts could be created and imported into COMSOL to simulate the cooling process. The cooling efficiency and improvement in the lens temperature distribution were compared between the GBA- and RGA-optimized channels.

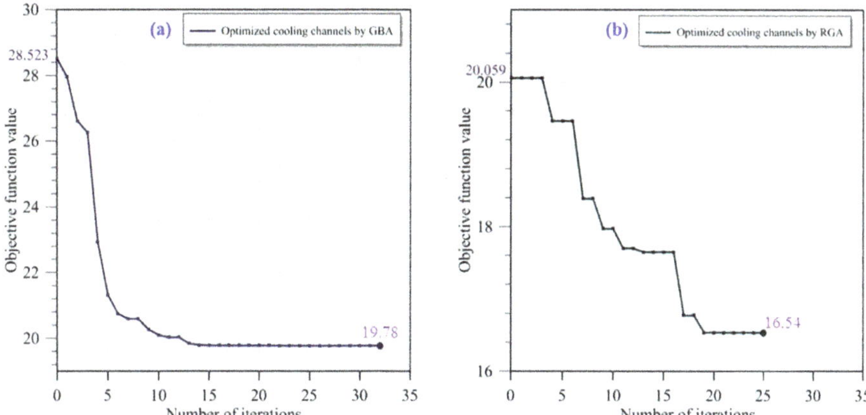

Figure 10. History of the objective function evolved from: (**a**) the GBA and (**b**) RGA during optimization.

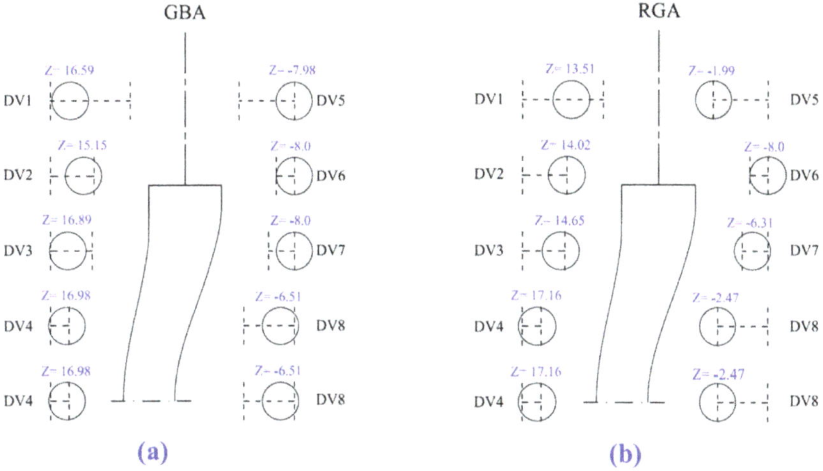

Figure 11. Optimized DVs obtained from: (**a**) the GBA and (**b**) RGA.

Table 2. The optimized DVs in Figure 11 were obtained from the GBA and RGA.

Optimized Parameters	DV1	DV2	DV3	DV4	DV5	DV6	DV7	DV8
Results of GBA	16.59	15.15	16.89	16.98	−7.98	−8.00	−8.00	−6.51
Results of RGA	13.51	14.02	14.65	17.16	−1.99	−8.00	−6.31	−2.47

The conformal cooling channels designed using the GBA were used to simulate the cooling process. The results indicated the variations in the maximum and minimum surface temperature of the melt in the mold cavity, as displayed in Figure 12a. For the planar conformal cooling channel, the default maximum surface temperature for the ejected lens was set as 359.713 K. For a temperature of 359.713 K, the ejection time for the GBA-optimized conformal cooling channels was 16.9 s. However, according to the cooling time suggested by Moldex3D, the lens was ejected at 18.6 s. The temperature distribution for the surface of the lens at 18.6 s is displayed in Figure 12b, with a maximum temperature of 358.906 K, a minimum temperature of 358.173 K, and a temperature difference of 0.733 K. Moreover, the ejection time for the RGA-optimized conformal cooling channels was 14.55 s (Figure 13a). The temperature distribution on the lens surface at 18.6 s for the RGA-optimized channels is displayed in Figure 13b, with a maximum temperature of 357.861 K, a minimum temperature of 357.255 K, and a temperature difference of 0.606 K. Thus, after the planar conformal cooling channels were modified, the GBA- and RGA-optimized three-dimensional conformal cooling channels enhanced the cooling efficiency and improved the temperature distribution of the melt in the mold cavity.

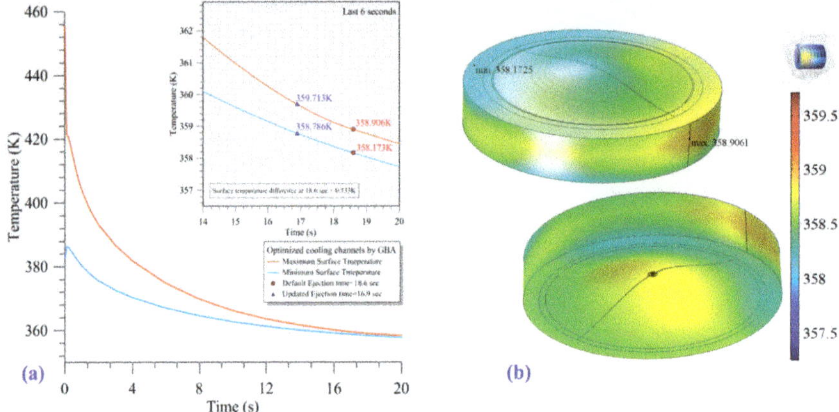

Figure 12. (**a**) Surface temperature of the melt during the cooling stage and (**b**) surface temperature of the melt at the default ejection time of 18.6 s when using the GBA-optimized cooling channels.

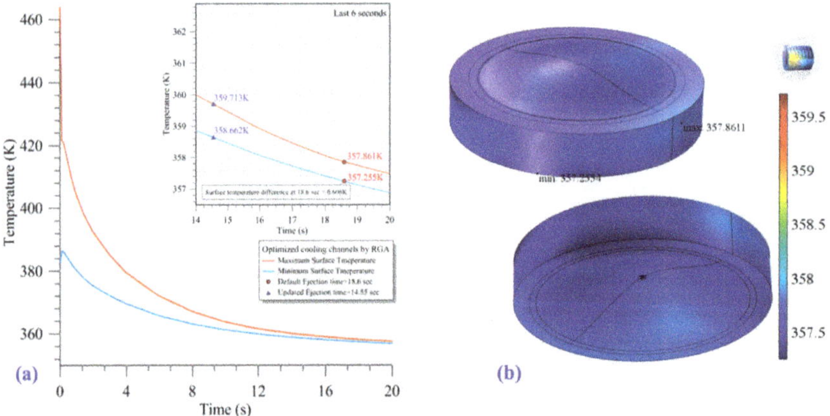

Figure 13. (**a**) Surface temperature of the melt during the cooling stage and (**b**) surface temperature of the melt at the default ejection time of 18.6 s when using the RGA-optimized cooling channels.

3.2. Comparison of the Conformal Cooling Channels Designed by Using Different Algorithms for the Entire Injection Molding Process

The conformal cooling channels designed using the GBA and RGA optimization algorithms were imported into Rhinoceros 3D software, which was used to construct the solid mesh models. The models were then used to execute the entire injection molding simulation with the Moldex3D software. The computation time was approximately 45 min when using two 2.4-GHz Intel Xeon E5-2620 CPUs with 64 GB of RAM. The results indicated that the optimized conformal cooling channels exhibited a lower average surface temperature for the melt compared with that exhibited by the planar conformal cooling channels (Figure 14). At the end of cooling, the surface temperatures simulated using Moldex3D were consistent with those simulated using COMSOL (Figure 9b, Figure 12b, and Figure 13b). If the temperature on the lens surface could be evenly distributed, the thermal-displacement-induced warpage could be improved. Therefore, to evaluate the thermally induced warpage, 18 measured nodes were selected on the lens surface. The locations and numbering of these measured nodes are displayed in Figure 15a. Measured nodes 1–16 were located on the top and bottom edges of the lens and used to calculate the total thermal warpage. The results were used to evaluate the variation in the circularity (roundness) due to the warpage. The average thermal warpage values for the planar conformal cooling channels, GBA-optimized conformal cooling channels, and RGA-optimized conformal cooling channels were 98.296, 98.214, and 98.121 µm, respectively (Figure 15b,c). Thus, the optimized conformal cooling channels exhibited a lower thermal warpage than the planar conformal cooling channels. Moreover, measured nodes 17 and 18 were located at the centers of the top and bottom surfaces of the lens, respectively. These two measured nodes were used to observe the relative thermal warpage in the vertical direction (z-direction). The vertical thermal warpage values for the GBA- and RGA-optimized conformal cooling channels were 21.617 and 21.565 µm, respectively. Both these values were lower than the vertical thermal warpage of the planar conformal cooling channels (21.638 µm).

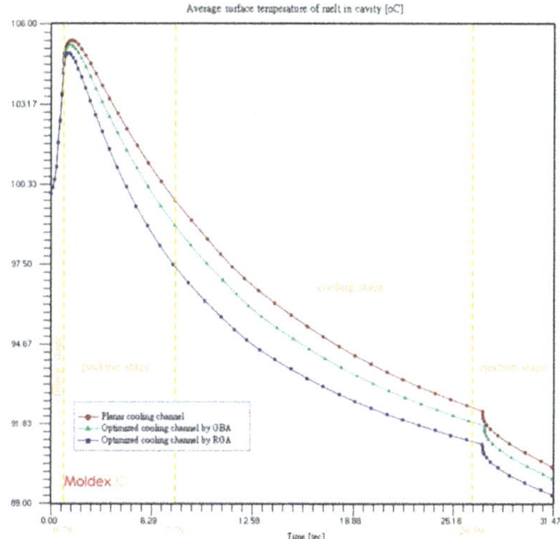

Figure 14. Average surface temperature of the melt in cavity during the complete injection molding cycle with different conformal cooling channels.

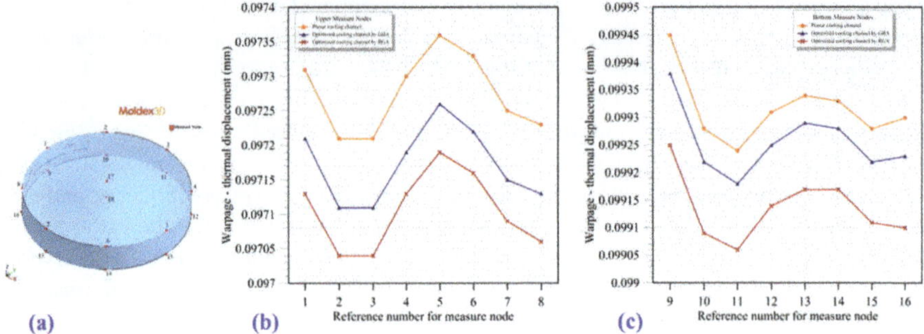

Figure 15. (**a**) Reference numbers and locations of the 18 measured nodes on the lens; (**b**) total thermal warpage values measured using the upper nodes (1–8); and (**c**) total thermal warpage values measured using the bottom nodes (9–16).

When designing cooling channels, the rules for their geometric layout should be considered and the temperature difference between the inlet and outlet should be minimized to prevent the warpage of the lens due to uneven temperature distribution in the mold. Generally, if the product requires high accuracy, the temperature difference between the inlet and outlet should be lower than 2.5 °C [28]. The simulation results of the GBA- and RGA-optimized cooling channels indicated that the temperature difference between the inlet and outlet of the optimized cooling channels was lower than that of the planar conformal cooling channels regardless of the upper and bottom channels (Figure 16). The GBA- and RGA-optimized conformal cooling channels improved the temperature uniformity of the mold.

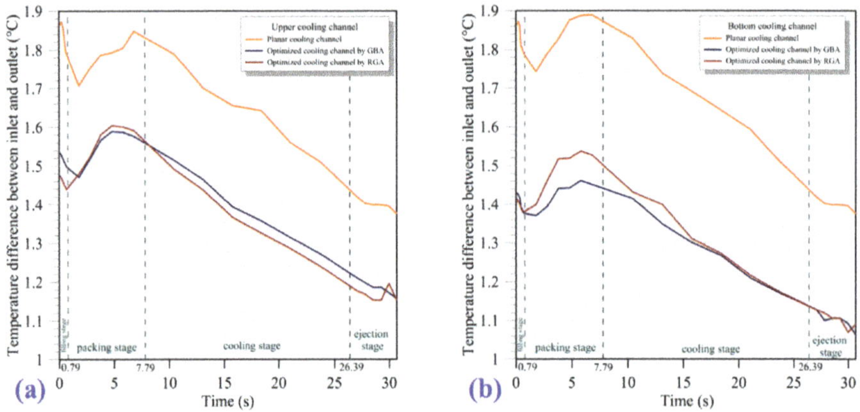

Figure 16. Temperature difference between the inlet and outlet for: (**a**) the upper cooling channel and (**b**) bottom cooling channel.

3.3. Optimal Process Parameters for Injection Molding

After the geometric optimization, the cooling channel with the highest cooling efficiency for the mold and the most even temperature distribution for the lens surface was adopted. Thus, the RGA-optimized cooling channel layout was selected for the DOE to determine the most suitable molding conditions. The optimal process parameters were selected on the basis of the signal/noise (S/N) ratio response. The optical properties of the lenses were compared before and after performing the DOE. Figure 17 illustrates the calculation results of the S/N ratio response from the DOE module of Moldex3D. The total S/N ratio response was obtained by adding the S/N ratio responses from three quality factors under four control factors and five levels (Figure 17d). A high S/N ratio indicates low noise (external influence). According to the Taguchi method, the level with the maximum S/N ratio was selected as the optimal condition [16,29]. Therefore, the optimal combination of control factors, reported in Table 1, and their corresponding levels was A4B2C4D1. The optimal control factor level settings were A4 (melt temperature: 285 °C), B2 (mold temperature: 89.75 °C), C4 (filling time: 0.99 s), and D1 (packing time: 6 s). This combination of process parameters (A4B2C4D1) was used in the injection molding simulation. The optics module of Moldex3D was adopted to predict the birefringence and fringe pattern of the lens. After the DOE analysis, the birefringence of the lens considerably decreased (Figure 18a), which indicated that the residual stress within the material also diminished [16]. Moreover, after the processing optimization, the fringe pattern on the lens resembled concentric circles (Figure 18b). The distributions of the fringe pattern on the optimized molded lens were rarefied near the gate, which implied that the optical properties of the lens had been improved.

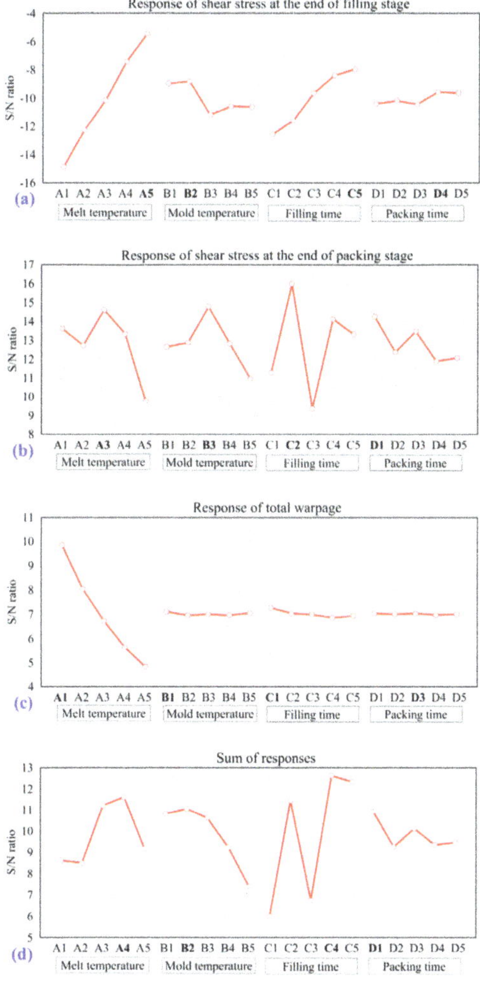

Figure 17. S/N ratios for: (**a**) shear stress at the end of the filling stage; (**b**) shear stress at the end of the packing stage; (**c**) total warpage; and (**d**) sum of responses from (**a**) to (**c**).

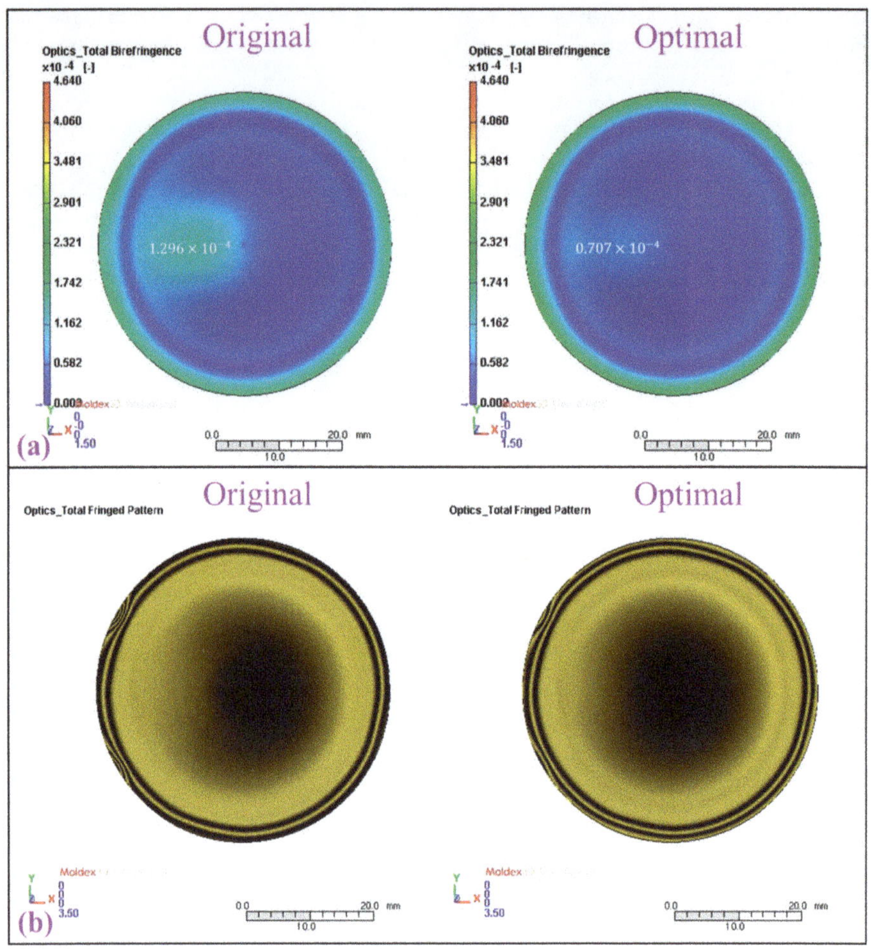

Figure 18. Comparison of the (**a**) birefringence properties and (**b**) fringe patterns before and after the optimization of process parameters.

4. Conclusions

In this study, the GBA and RGA optimization algorithms were used to design the geometric layout of conformal cooling channels. The planar conformal cooling channels had a lens surface temperature difference of 1.112 K with an ejection time of 18.6 s. The GBA-optimized conformal cooling channels reduced the ejection time to 16.9 s and improved the cooling efficiency by approximately 9.14%. The RGA-optimized conformal cooling channels had an ejection time of 14.55 s and improved the cooling efficiency by approximately 21.77%. The GBA-optimized conformal cooling channels exhibited a decrease in the temperature difference on the lens surface from 1.112 to 0.733 K, which was an improvement of 34.08%. Furthermore, the RGA-optimized conformal cooling channel exhibited a decrease in the temperature difference on lens surface from 1.112 to 0.606 K, which was an improvement of 45.5%.

During the entire injection molding cycle, the planar conformal cooling channels exhibited the highest average surface temperature of the melt, followed by the GBA-optimized conformal cooling channels. The RGA-optimized conformal cooling channels exhibited the lowest average surface temperature of the melt. Furthermore, the distribution of the thermal-displacement-induced warpage of the lens could be used as an index to evaluate whether the cooling channel system was suitably designed. The GBA- and RGA-optimized conformal cooling channels exhibited improvements of 0.18% and 0.52%, respectively, in average thermal warpage of measured nodes as compared with the planar conformal cooling channels. However, thermally induced warpage still could not be compared with experimental results because of insufficiently available experimental data.

In summary, the RGA-optimized conformal cooling channels had the highest cooling efficiency for the mold and a superior temperature distribution for the melt in the mold cavity, which reduced thermally induced warpage. Therefore, using RGA-optimized conformal cooling channels shortened the development time and considerably improved the quality of the plastic lens. DOE was used to evaluate the effect of manufacturing process parameters on the optical properties. The optimal process parameters reduced the birefringence and improved the shape of the fringe pattern.

Recently, three-dimensional printing and additive manufacturing have attracted tremendous attention worldwide. Among these technologies, the additive manufacturing of metal powder has quickly developed into a feasible technique for fabricating metal parts. Additive manufacturing typically uses a high-energy electron beam or laser beam to sinter or melt the metal powder to form the solid parts [30,31]. This method can be used to fabricate mold cores for plastic product manufacturing and construct three-dimensional conformal cooling channels [32,33]. The methods presented in this study can serve as useful references to obtain optimized cooling channel layouts and injection molding conditions for the initial stages of mold development. Mold tooling with conformal cooling channels can be used to improve the cooling efficiency and temperature distribution of the melt, which solves warpage issues. Optimal injection molding conditions can improve product quality. By controlling key molding conditions, the product yield rate can be enhanced, and the product cost can be decreased.

Supplementary Materials: The following are available online at http://www.mdpi.com/2076-3417/9/20/4341/s1. The integration of the COMSOL finite element model and SmartDO genetic optimization algorithm is available in the online version.

Author Contributions: Conceptualization, C.-Y.C.; methodology, C.-Y.C.; formal analysis, C.-Y.C.; investigation, C.-Y.C.; data curation, C.-Y.C.; writing—original draft preparation, C.-Y.C.; writing—review and editing, C.-Y.C.; supervision, C.-Y.C.; project administration, C.-Y.C.; funding acquisition, C.-Y.C.

Funding: This research was funded by the Ministry of Science and Technology in Taiwan under the grant number MOST 104-2218-E-008-006.

Acknowledgments: The author would like to thank his student, Po-Yang Chang, was of great help in generating some of the pictures shown in this paper. Special thanks for manager of EOS Singapore, Stephanie Cheong, providing an approval of reproducing Figure 4 with permission of the copyright owner. Also, thanks for Shih-Ying Ke from Glory Science Company providing the AutoCAD file of Figure 1.

Conflicts of Interest: The finite element software COMSOL was run under the license number 5084742. The author has no conflicts of interest related to this manuscript, commercial or otherwise.

References

1. Garrette, B.; Karnani, A. Challenges in marketing socially useful goods to the poor. *Calif. Manag. Rev.* **2010**, *52*, 29–47. [CrossRef]
2. Kohara, T. Development of new cyclic olefin polymers for optical uses. *Macromol. Symp.* **1996**, *101*, 571–579. [CrossRef]
3. Lee, Y.B.; Kwon, T.H. Modeling and numerical of residual stresses and birefringence in injection molded center-gated disks. *J. Mater. Process. Technol.* **2001**, *111*, 214–218. [CrossRef]
4. AC Technology Inc. *C-MOLD Design Guide: A Resource for Plastics Engineers*; Advanced CAE Technology: Ithaca, NY, USA, 1995.
5. Sachs, E.; Wylonis, E.; Allen, S.; Cima, M.; Guo, H. Production of injection molding tooling with conformal cooling channels using the three dimensional printing process. *Polym. Eng. Sci.* **2000**, *40*, 1232–1247. [CrossRef]

6. Dalgarno, K.W.; Stewart, T.D. Manufacture of production injection mould tooling incorporating conformal cooling channels via indirect selective laser sintering. *Proc. Inst. Mech. Eng. Part B* **2001**, *215*, 1323–1332. [CrossRef]

7. Yan, C.; Hsu, A. Introduction of Composite Technology, Combining Machining with Selective Laser Melting for Metal Powder Forming. In *Molding Innovation Newsletter Feb. 2012*; CoreTech System Co., Ltd.: Chupei City, Taiwan, 2012; pp. 5–10.

8. Hsu, F.H.; Wang, K.; Huang, C.T.; Chang, R.Y. Investigation on conformal cooling system design in injection molding. *Adv. Prod. Eng. Manag.* **2013**, *8*, 107–115. [CrossRef]

9. Qiao, H. A systematic computer-aided approach to cooling system optimal design in plastic injection molding. *Int. J. Mech. Sci.* **2006**, *48*, 430–439. [CrossRef]

10. Park, H.-S.; Dang, X.-P. Optimization of conformal cooling channels with array of baffles for plastic injection mold. *Int. J. Precis. Eng. Manuf.* **2010**, *11*, 879–890. [CrossRef]

11. Dang, X.-P.; Park, H.-S. Design of U-shape milled groove conformal cooling channels for plastic injection mold. *Int. J. Precis. Eng. Manuf.* **2011**, *12*, 73–84. [CrossRef]

12. Jahan, S.A.; Wu, T.; Zhang, Y.; Zhang, J.; Tovar, A.; El-Mounayri, H. Thermo-mechanical design optimization of conformal cooling channels using design of experiments approach. *Procedia Manuf.* **2017**, *10*, 898–911. [CrossRef]

13. Li, Z.; Wang, X.; Gu, J.; Ruan, S.; Shen, C.; Lyu, Y.; Zhao, Y. Topology optimization for the design of conformal cooling system in thin-wall injection molding based on BEM. *Int. J. Adv. Manuf. Technol.* **2018**, *94*, 1041–1059. [CrossRef]

14. Jahan, S.A.; El-Mounayri, H. A thermomechanical analysis of conformal cooling channels in 3D printed plastic injection molds. *Appl. Sci.* **2018**, *8*, 2567. [CrossRef]

15. Chen, C.-C.A.; Vu, L.T.; Qiu, Y.-T. Study on injection molding of shell mold for aspheric contact lens fabrication. *Procedia Eng.* **2017**, *184*, 344–349. [CrossRef]

16. Lin, C.-M.; Hsieh, H.-K. Processing optimization of Fresnel lenses manufacturing in the injection molding considering birefringence effect. *Microsyst. Technol.* **2017**, *23*, 5689–5695. [CrossRef]

17. Shieh, J.-Y.; Wang, L.K.; Ke, S.-Y. A feasible injection molding technique for the manufacturing of large diameter aspheric plastic lenses. *Opt. Rev.* **2010**, *17*, 399–403. [CrossRef]

18. Chen, C.-C.A.; Tang, J.-C.; Teng, L.-M. Effects of mold design of aspheric projector lens for head up display. In Proceedings of the Polymer Optics Design, Fabrication, and Materials, San Diego, CA, USA, 12 August 2010; Volume 7788, p. 778806.

19. Mayer, S. *Optimised Mould Temperature Control Procedure Using DMLS*; EOS Whitepaper 2005; EOS GmbH Ltd.: Krailling, Germany, 2005; pp. 1–10.

20. G-Plast Pvt. Ltd. Synergy of True & Full 3D Simulation and Conformal Cooling. In *Molding Innovation Newsletter Feb. 2012*; CoreTech System Co., Ltd.: Chupei City, Taiwan, 2012; pp. 18–20.

21. Wang, P.-J.; Lai, H.-E. Study of residual birefringence in injection molded lenses. In *Annual Technical Conference (ANTEC)*; SPE- the Society of Plastics Engineers: Cincinnati, OH, USA, 2007; pp. 2494–2498.

22. Lai, H.-E.; Wang, P.-J. Study of process parameters on optical qualities for injection-molded plastic lenses. *Appl. Opt.* **2008**, *47*, 2017–2027. [CrossRef]

23. COMSOL Inc. Application ID: 12371, Cooling of an Injection Mold. Available online: http://www.comsol.com/model/12371 (accessed on 1 July 2016).

24. Chung, C.-Y.; Mansour, J.M. Determination of poroelastic properties of cartilage using constrained optimization coupled with finite element analysis. *J. Mech. Behav. Biomed. Mater.* **2015**, *42*, 10–18. [CrossRef]

25. Chen, Y.-C.; Huang, B.-K.; You, Z.-T.; Chan, C.-Y.; Huang, T.-M. Optimization of lightweight structure and supporting bipod flexure for a space mirror. *Appl. Opt.* **2016**, *55*, 10382–10391. [CrossRef]

26. Chen, S.-Y. An approach for impact structure optimization using the robust genetic algorithm. *Finite Elem. Anal. Des.* **2001**, *37*, 431–446. [CrossRef]

27. Lin, C.-M.; Wang, C.-K. Processing optimization of optical lens in the injection molding. *Adv. Mater. Res.* **2013**, *813*, 161–164. [CrossRef]

28. Himasekhar, K.; Lottey, J.; Wang, K.K. CAE of mold cooling in injection molding using a three-dimensional numerical simulation. *J. Eng. Ind.* **1992**, *114*, 213–221. [CrossRef]

29. Chen, D.-C.; Huang, C.-K. Study of injection molding warpage using analytic hierarchy process and Taguchi method. *Adv. Technol. Innov.* **2016**, *1*, 46–49.
30. Mazur, M.; Brincat, P.; Leary, M.; Brandt, M. Numerical and experimental evaluation of a conformally cooled H13 steel injection mould manufactured with selective laser melting. *Int. J. Adv. Manuf. Technol.* **2017**, *93*, 881–900. [CrossRef]
31. Abbès, B.; Abbès, F.; Abdessalam, H.; Upganlawar, A. Finite element cooling simulations of conformal cooling hybrid injection molding tools manufactured by selective laser melting. *Int. J. Adv. Manuf. Technol.* **2019**, *103*, 2515–2522. [CrossRef]
32. Kitayama, S.; Miyakawa, H.; Takano, M.; Aiba, S. Multi-objective optimization of injection molding process parameters for short cycle time and warpage reduction using conformal cooling channel. *Int. J. Adv. Manuf. Technol.* **2017**, *88*, 1735–1744. [CrossRef]
33. Kuo, C.-C.; Jiang, Z.-F. Numerical and experimental investigations of a conformally cooled maraging steel injection molding tool fabricated by direct metal printing. *Int. J. Adv. Manuf. Technol.* **2019**. [CrossRef]

Article

An Entropy Weight-Based Lower Confidence Bounding Optimization Approach for Engineering Product Design

Jiachang Qian [1,2], Jiaxiang Yi [1], Jinlan Zhang [2], Yuansheng Cheng [1,3] and Jun Liu [1,3,*]

[1] School of Naval Architecture and Ocean Engineering, Huazhong University of Science and Technology, Wuhan 430074, China; qianjiachang@163.com (J.Q.); jiaxiangyi@hust.edu.cn (J.Y.); yscheng@hust.edu.cn (Y.C.)
[2] Wuhan Second Ship Design and Research Institute, Wuhan 430064, China; zjlwh719@sina.com
[3] Collaborative Innovation Center for Advanced Ship and Deep-Sea Exploration (CISSE), Shanghai 200240, China
* Correspondence: hustlj@hust.edu.cn

Received: 15 January 2020; Accepted: 18 May 2020; Published: 21 May 2020

Abstract: The optimization design of engineering products involving computationally expensive simulation is usually a time-consuming or even prohibitive process. As a promising way to relieve computational burden, adaptive Kriging-based design optimization (AKBDO) methods have been widely adopted due to their excellent ability for global optimization under limited computational resource. In this paper, an entropy weight-based lower confidence bounding approach (EW-LCB) is developed to objectively make a trade-off between the global exploration and the local exploitation in the adaptive optimization process. In EW-LCB, entropy theory is used to measure the degree of the variation of the predicted value and variance of the Kriging model, respectively. Then, an entropy weight function is proposed to allocate the weights of exploration and exploitation objectively and adaptively based on the values of information entropy. Besides, an index factor is defined to avoid the sequential process falling into the local regions, which is associated with the frequencies of the current optimal solution. To demonstrate the effectiveness of the proposed EW- LCB method, several numerical examples with different dimensions and complexities and the lightweight optimization design problem of an underwater vehicle base are utilized. Results show that the proposed approach is competitive compared with state-of-the-art AKBDO methods considering accuracy, efficiency, and robustness.

Keywords: Kriging; lower confidence bounding; entropy theory; product design; simulation-based design optimization

1. Introduction

Computational simulation models, i.e., finite element analysis (FEA) and computational fluid dynamic (CFD) models, have been widely used in engineering design problems to replace physical experiments for reducing the time cost and shortening the product developing cycle. However, it is still computationally prohibited to solve engineering design optimization problems directly relying on simulation models, even though the storage capacity and computing efficiency of computers are maintaining rapid growth [1,2]. A popular strategy to address this limitation is to adopt surrogate models, also named the meta-model or approximate model, to replace the computational simulation model during the optimization process. There are several varieties of surrogate models, such as Polynomial response surface (PRS) model [3], Radial basis function (RBF) model [4,5], Kriging model [6–8], and Support vector regression (SVR) model [9,10]. Among these surrogate models, the

Kriging model has been intensively used in engineering design optimization because it can provide not only the predicted value of an un-sampled point but also the predicted confidence interval of the predicted value.

The Kriging model-based design optimization methods can be divided into two types [11]: the off-line type and the on-line type. The off-line type uses all the computational resources to construct the final Kriging model in one time and the model does not update during the optimization process. Therefore, the determination of the sampling plan is crucial because the optimization process may fall into the local optimum region if the number of samples is too small [12]. On the contrary, it would waste computational burden if the number of samples is too large. To solve this dilemma, the on-line type has been developed, in which an initial Kriging model is built in the earlier stage and then new samples are added to update the Kriging model sequentially through certain criteria, e.g., maximum mean square error [13], cross validate error [14], etc., during the design optimization process. The on-line type can significantly reduce the computational burden compared with that of the off-line type because the information of the Kriging model is well utilized [15,16].

The on-line type Kriging model-based design optimization methods are also called adaptive Kriging-based design optimization (AKBDO). The target of AKBDO is to obtain the optimum using less computational cost [17,18]. At the same time, the balance between exploration and exploitation is important because it is critical for searching the global optimum. In detail, exploration means the ability of the algorithm to explore the whole design space for the latent optimal region. On the other hand, exploitation aims to identify the local area around the current optimum. Typically, there are several sorts of adaptive sampling approaches of AKBDO with different ways of making a trade-off between exploration and exploitation [19,20], such as the maximum-uncertainty adaptive sampling approaches [20], the efficient global optimization (EGO) methods [21], the lower confidence bounding (LCB) based methods [22], the aggregate-criteria adaptive sampling methods [19], and the multi-criteria adaptive approaches [23,24]. Among these approaches, the efficient global optimization method proposed by Jones [21] has been intensively adopted to handle realistic product design due to its high efficiency and ease of operation. In this work, the expected improvement (EI) function is introduced to quantify the improvement of an un-known point to the current best solution. The new point can be obtained by maximizing the EI function, and the Kriging model can be updated adaptively by adding the new point to the original sample set. The EI based EGO method has been intensively investigated in recent years [25–27]. For example, Xiao [28] proposed a weighted EI to make the balance between exploration and the exploitation more flexible; Zhan [29] proposed the EI matrix method to solve the multi-objective problem. Another famous AKDBO method is the LCB- based method [30]. The LCB function is an effective approach to balance exploration and exploitation by combining the predicted value and variance in a simple way [31]. Subsequently, a parameterized LCB (PLCB) method was proposed by using cool strategy to improve the ability to balance the exploration and exploitation of the original LCB [32]. Cheng et al. [33] considered the coefficient of variation of predicted values and variance to determine the weight factor adaptively during the sequential process. Further, some variants of the LCB methods focus on upper confidence bounding, such as, the Gaussian process upper confidence bounding (GPUCB) algorithm proposed by Srinivas et al., which considers the upper confidence bound of noisy functions [34]. The parallel type of the GPUCB was developed by Desautel et al. [35]. It mentions that LCB methods have been widely applied to solve real engineering problems [36,37]. However, the weight factor for balancing the exploration and the exploitation of the LCB based method remains an interesting problem. This is because most of the existing factor approaches are subjective or problem dependent, which is not robust in application for all cases.

In this paper, an entropy weight-based lower confidence bounding approach (EW-LCB) is proposed to ascertain the weight of the LCB function adaptively and objectively. In the proposed EW-LCB method, entropy theory is used to quantify the degree of variation of the predicted value and variance of the Kriging model. Then, a new weighted formula is introduced to allocate the weights of exploration and exploitation adaptively. To validate the performance of the proposed EW- LCB method, several numerical functions with different dimensions and complexities and an engineering problem are tested. The computational efficiency, accuracy of the optimum, and the robustness are considered when comparing EW-LCB with the existing famous AKBDO methods. Results showed that the performances of the proposed EW-LCB approach were competitive on the test cases.

The remainder of this paper is organized as follows, in Section 2, the basis of the Kriging model and several existing famous AKBDO methods are introduced. The details of the proposed approach with the assistance of an illustrative example are described in Section 3. In Section 4, the effectiveness of the proposed approach is tested on several numerical benchmark problems and an engineering design optimization problem. Finally, some conclusions and possible future works are proüposed in Section 5.

2. Background

2.1. Kriging Model

The Kriging model was originally proposed by Krige [38] to predict the location of a mine hole in a geostatistical community. Then, it was extended by Sacks et al. [6] for modeling an experiment of a computer. The Kriging model is also called the Gaussian process model, which is a kind of interpolative model. The Kriging model can be expressed as

$$\hat{y}(x) = \beta + Z(x) \tag{1}$$

where x represents the vector of the design variables, which is a d-dimensional vector $x = \{x_1, x_2, \ldots, x_d\}$, β is an unknown parameter which denotes the global tendency, $Z()$ is a static Gaussian process with zero mean and non-zero variance σ^2, which represents the local deviation.

In the static Gaussian process, spatial correlation is used to organize the relationship between any two samples. Generally, the squared exponential function is utilized, which can be expressed as

$$R(x_i, x_j; \theta) = \exp(-\sum_{k=1}^{d} \theta_k (x_i^k - x_j^k)^{p_j}) \tag{2}$$

where θ and P are the hyper-parameters used to control the smooth and the correlation between two sample points. Generally, the hyper-parameter vector P is set to be $p_i = 2; i = 1, 2, \ldots, d$ [39].

The core point of the modeling process of the Kriging model is to determine the unknown parameters. Because the responses obey the multivariable Gaussian distribution, the unknown parameter can be obtained by maximum likelihood estimation (MLE) [14]. The likelihood function can be organized as

$$L\big(y(x_1), y(x_2), \ldots, (x_N)\big|\sigma, \beta, \theta\big) = \frac{1}{(2\pi\sigma^2)^{\frac{N}{2}}} \exp\left[-\frac{\sum\limits_{i=1}^{N} (y(x_i) - \beta)^2}{2\sigma^2}\right] \tag{3}$$

where N is the number of samples.

Then, Equation (3) can be simplified by taking the natural logarithm,

$$\ln(L) = -\frac{N}{2}\ln(2\pi) - \frac{N}{2}\ln(\sigma^2) - \frac{1}{2}\ln(|R|) - \frac{(y - 1\beta)^T R^{-1}(y - 1\beta)}{2\sigma^2} \tag{4}$$

where y is an N-dimensional vector that consists of the real responses, 1 is an N-dimensional vector that consists of 1,

The values of β and σ^2 can be obtained by setting the derivatives of Equation (4) concerning β and σ^2 to be 0,

$$\hat{\beta} = \frac{f^T R^{-1} y}{f^T R^{-1} f} \tag{5}$$

$$\hat{\sigma}^2 = \frac{\left(y - 1\hat{\beta}\right)^T R^{-1}\left(y - 1\hat{\beta}\right)}{N} \tag{6}$$

Then, substituting Equations (5) and (6) into Equation (4), and remove the constant terms, Equation (4) yields the concentrated ln-likelihood function

$$\ln(L) = -\frac{N}{2}\ln\left(\hat{\sigma}^2\right) - \frac{1}{2}\ln(|R|) \tag{7}$$

It is difficult to obtain an analytical solution of θ because of high non-linearity and non-differentiality. Therefore, a numerical solution is obtained instead. The optimization algorithm, such as the genetic algorithm (GA) [40] and particle swarm optimization algorithm (PSO) [41], can be used to find the optimized values of θ.

The Kriging model is widely adopted in surrogate model-based engineering optimization because it can provide both the predicted value and variance [42]. The predicted value of an un-sampled point can be determined by minimizing the mean square error. Thus, the predicted value and variance can be expressed as

$$\hat{y}(x) = \hat{\beta} + r(x)^T R^{-1}\left(y - 1\hat{\beta}\right) \tag{8}$$

$$\hat{s}^2(x) = \hat{\sigma}^2\left[1 - r(x)^T R^{-1} r(x) + \frac{\left(1 - 1^T R^{-1} r(x)\right)^2}{1^T R^{-1} 1}\right] \tag{9}$$

where $r(x)$ is an N-dimensional vector representing the spatial correlation between the un-sample point and the sample points, which can be defined by

$$r(x) = \{R(x, x_1), R(x, x_2), \dots, R(x, x_N)\} \tag{10}$$

2.2. Review of the Typical Adaptive Surrogate-Based Design Optimization Methods

The goal of the AKBDO methods is to obtain the optimum with a limited computational budget. In this section, four popular AKBDO methods are briefly introduced.

2.2.1. The Lower Confidence Bounding Method

With a concise expression, the LCB method is a popular AKBDO method, which can be expressed as

$$lcb(x) = \hat{y}(x) - b\hat{s}(x) \tag{11}$$

where $\hat{y}(x)$ and $\hat{s}(x)$ are the predicted value and standard deviation, respectively. b is a factor utilized to control the weight between the $\hat{y}(x)$ and $\hat{s}(x)$ for the sake of balancing the exploration and the exploitation.

The goal of the LCB function is to identify the new sample points through the combination of predicted value and variance by Equation (11). The point with small predicted value or large uncertainty is chosen. Generally, a larger b means more emphasis on global exploration. On the contrary, with a small b value, the algorithm turns more attention to local exploitation. Cox and John reported that $b = 2$ and $b = 2.5$ can give a more efficient search [43].

2.2.2. The Parameterized Lower Confidence Bounding Method

The weight factor in the LCB method is constant, indicating that the contributions of the predicted value and standard deviation will be fixed during the optimization process. Thus, the parameterized lower confidence bounding (PLCB) method is proposed [32], which can be defined by

$$plcb(x) = a_i\hat{y}(x) - b_i\hat{s}(x) \tag{12}$$

where a new parameter a_i is developed to regulate the influence of the predicted value during the iteration process of design optimization. Meanwhile, the values of a_i and b_i vary during the iteration process, where i is the iteration order of the sequential process. In detail, the values of the parameters a_i and b_i can be expressed as

$$a_i = 1, b_i = \left(1 + \cos\left(\frac{i\pi}{m}\right)\right) / \sin\left(\frac{i\pi}{m}\right) \tag{13}$$

where m is a parameter defined by the user, it is set to be $m = 3$ in Ref. [32].

According to Equation (12), the algorithm tends to focus on exploration when b_i/a_i has a larger value, while it tends to focus on exploitation when b_i/a_i has a respective small value. Specifically, the value of b_i/a_i in PLCB function has a larger value at the former iterations and has a relatively small value as the algorithm goes on. Consequently, the PLCB algorithm shows a better ability to balance the exploitation and the exploration when compared with the LCB method.

2.2.3. The Expected Improvement Method

The expected improvement method is a famous AKBDO method proposed by Jones [21]. The expected improvement function can be defined to measure the latent improvement of an unknown point to the current optimum, which can be expressed as

$$I(x) = \max(y_{\min} - Y(x), 0) \tag{14}$$

The expected improvement can be formalized as

$$E(I(x)) = E(\max(y_{\min} - Y(x), 0)) \tag{15}$$

which can be expanded into

$$E[I(x)] = (f_{\min} - \hat{y}(x))\Phi\left(\frac{f_{\min} - \hat{y}(x)}{s(x)}\right) + \hat{s}(x)\phi\left(\frac{f_{\min} - \hat{y}(x)}{\hat{s}(x)}\right) \tag{16}$$

where Φ and ϕ are the cumulative density function and probability density function of the standard normal distribution, respectively.

According to Equation (16), the first term mainly focuses on the exploitation and the second term primarily concerns the exploration. The point with the maximum value of the EI function is regarded as the new sample to update the Kriging model during the iteration process.

2.2.4. The Weighted Expected Improvement Method

Although the EI method can balance the exploration and the exploitation, its efficiency is problem-dependent because the EI method provides a fixed compromise between the exploration and the exploitation. To address this issue, a weighted expected improvement method (WEI) [28] is developed, in which a tunable weight is adopted to adjust the contributions of exploration and exploitation. The WEI can be given by

$$E[I(x)] = w(f_{\min} - \hat{y}(x))\Phi\left(\frac{f_{\min} - \hat{y}(x)}{s(x)}\right) + (1-w)\hat{s}(x)\phi\left(\frac{f_{\min} - \hat{y}(x)}{\hat{s}(x)}\right) \tag{17}$$

where w is the weight coefficient. The larger value of w indicates that the WEI will focus more on exploitation. Otherwise, the WEI method emphasizes exploration.

3. Proposed Approach

The goal of the proposed lower confidence bounding approach based on the entropy weight algorithm (EW-LCB) is to obtain an optimal solution with less computational burden through a sequential process. In EW-LCB, a new-weight factor is developed, which can allocate factors to balance global exploration and local exploitation by quantifying the degree of variation of the predicted value and variance from the Kriging model, respectively. In detail, the entropy theory is adopted to evaluate the relative discrepancy between the predicted value and uncertainty of the Kriging model. The framework of the EW-LCB is shown in Figure 1, which is composed of six steps.

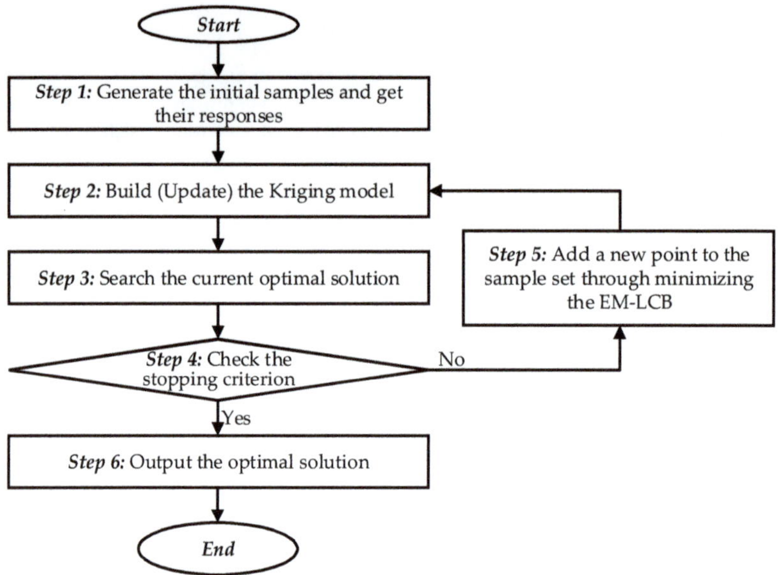

Figure 1. The framework of the proposed entropy weight algorithm (EW-LCB).

To demonstrate the proposed EW-LCB approach more intuitively and detailed, a one- dimensional toy example is utilized. The test function is adopted from [33], which can be expressed as

$$y = 0.5\sin(4\pi\sin(x+0.5)) + \frac{1}{3}(x+0.5)^2; \quad x \in [0,1] \tag{18}$$

The objective is to obtain the minimum value of Equation (18). Meanwhile, this function has a local optimal value $y = -0.0445$ at $x = 0$ and a global optimal value $y = -0.1341$ at $x = 0.5312$.

The details of the steps are elaborated as follows:

3.1. Step 1: Generate the Initial Sample Set

The generation of the initial sample set includes the determination of the number and location of the initial sample points, which is a crucial component of the AKBDO. If too few points are generated, the AKBDO can have a risk of falling into the local optimal because of the poor accuracy of the initial Kriging model. On the other side, it may be a waste of computational burden if too many initial samples are utilized, especially when dealing with costly engineering problems. For the tested cases, a state-of-the-art initial sample size rule $N = 10 \times d$ is used [21,44]. The sensitive analysis of the initial sample size is discussed in the next section. Besides, how to allocate the locations of the initial samples is another tricky issue. More uniformed distributed sample points are preferred because the initial Kriging model can obtain more information about the landscape of the real function. Therefore, the Latin Hypercube sampling (LHS) method [45] is used, which can guarantee that the samples distribute along each dimension uniformly.

Due to the simple landscape of the illustration example, the initial sample points are set to be $x = [0, 0.5, 1]$, which is less than the recommended initial sample size. Herein, the responses of the initial sample points are $y = [-0.0445, -0.1229, 0.7343]$, which are obtained by calculating the numerical function in Equation (18).

3.2. Steps 2 and 3: Constructing the Kriging Model and Obtaining the Current Optimal Solution

In Step 2, the Kriging model is established based on the initial sample set based on the DACE toolbox [46]. In detail, the regression function, the correlation function, and the initial value of θ are set to be '*Regpoly0*', '*Corrgauss*', and $(10d)^{-1/d}$, respectively. Besides, all the codes are executed based on the computational platform with a 4.2 GHz Intel(R) Eight-Core (TM) i7-7700k Processor and 64 GB RAM. The initial Kriging model of the illustrated example is plotted in Figure 2 in which, the black line and blue dash line denote the real function and the initial Kriging model, respectively. Meanwhile, the initial sample points are marked with blue triangles.

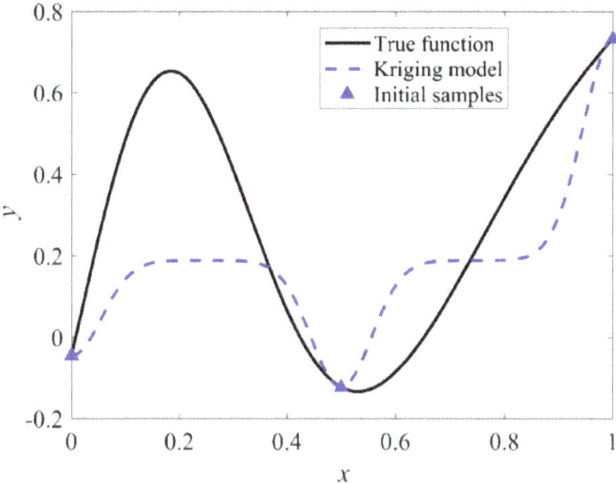

Figure 2. The initial Kriging model of the illustrated example.

In Step 3, the current optimal value was obtained through a genetic algorithm [47], where the parameter setting is listed in Table 1.

Table 1. The parameter setting of the genetic algorithm.

Parameter	Values
Population Size	100
Maximum generation	100
Crossover probability	0.95
Mutation probability	0.01

The minimum value of the current responses is −0.1229, which is larger than the actual global optimal solution. Then, the current minimum value will be judged by the stopping criterion to decide whether the active-learning process goes on or not in the next step.

3.3. Step 4: Check the Terminal Condition

Generally, there are two common ways to stop the sequential process. That is (1) the difference between the current optimal solution and the actual one achieves at an acceptable level and (2) all the computational resources are used up. In this work, these stopping criteria are adopted for different scenarios. For the numerical functions, because the actual optimal solution is known, the stopping criterion can be associated with this value to test the effectiveness of the proposed approach. Therefore, the stop condition is defined as

$$\varepsilon_r = \left| \frac{\min(y_k(x_i)) - y^r}{y^r} \right| < \varepsilon_g \; i = 1, 2, \cdots, N \tag{19}$$

where $\min(y_k(x_i))$ is the minimum value of the current sample set, y^r is the actual optimal solution, ε_g is a user-defined tolerance. Generally, the adaptive algorithm will confront the stricter test in the case of smaller tolerance. In this work, the value of ε_g is defined as 0.002 referring to [32].

However, for the engineering cases, the above stopping criterion for the numerical problem is impractical because the engineering problem is always a black-box problem. Thus the value of the actual optimal solution is unknown. Therefore, the sequential updating process terminates when the maximum iteration is reached, which can be expressed as

$$k \geq K \tag{20}$$

where k and K denote the current iteration and the maximum iteration, respectively.

If the stopping criterion is satisfied, the sequential process will be terminated and the algorithm goes to Step 6. Otherwise, the proposed algorithm goes to Step 5 for a new iteration. In this illustrated example, the relative error between the current optimal solution and the actual one is 0.00084. Therefore, the sequential process goes to Step 5.

3.4. Steps 5: Update the Sample Set through the Proposed EW-LCB

To accelerate the adaptive optimization process, the lower confidence bounding function based on entropy theory is developed. Entropy theory was proposed by Shannon to quantify the degree of chaos in molecular motion [48,49]. In this work, it is developed to quantify the degree of variation of the predicted value and variance in the sequential optimization process. Herein, the proposed entropy weight method is an objective weighting method, which adaptively assigns weight to the LCB function according to the degree of variation of the predicted value and variances. Specifically, the entropy weight method consists of three major steps: normalize the values of the predicted value and variances, calculate the entropy value of the predicted value and variances, and determine the relative weight of them.

The EW-LCB function is defined as

$$EWMLCB(x) = w_1 \hat{y}(x) - w_2 \hat{s}(x) \exp((-1)^r r) \tag{21}$$

where $\hat{y}(x), \hat{s}(x)$ are the predicted value and estimated standard deviation of the tested point, respectively. w_1, w_2 are the weights to reflect the contribution of the $\hat{y}(x), \hat{s}(x)$, respectively. r represents the iterations of the current optimization solution, which can be used to avoid the proposed approach falling in the local optimal region.

To obtain the weights w_1, w_2, suppose that there are N samples with m indexes. The information of the samples can be normalized by

$$Y_{ij} = \frac{X_{ij} - \min\{X_{1j}, X_{2j}, \dots, X_{Nj}\}}{\max\{X_{1j}, X_{2j}, \dots, X_{Nj}\} - \min\{X_{1j}, X_{2j}, \dots, X_{Nj}\}} \tag{22}$$

where X_{ij} represents the jth index of the ith sample. Equation (22) is used to normalize the lower and upper bound. In this work, the value of m equals 2. Besides, the number of tested points is set to be 1000 to improve the robustness of the entropy weight method.

Then, the entropy value of each index can be determined by

$$E(p_j) = -\frac{1}{\ln(N)} \sum_{i=1}^{N} p_{ij} \ln(p_{ij}) \; j = 1, 2, \cdots, m \tag{23}$$

where

$$p_{ij} = Y_{ij} / \sum_{i=1}^{N} Y_{ij} \tag{24}$$

If the value of $p_{ij} = 0$, it indicates that the entropy of this tested point equals zero. In that case, a definition is given to compensate for the insufficiency of the initial assumption in Equation (23), which is defined as

$$\lim_{p_{ij} \to 0} p_{ij} \ln(p_{ij}) = 0 \tag{25}$$

According to Equation (23), the degree of variation of each indicator can be ascertained. The indicator with a larger value of information entropy has a smaller degree of variation. Subsequently, the corresponding entropy weight should be small. As such, the entropy weight can be obtained by

$$w_j = \frac{1 - E(P_j)}{\sum\limits_{j=1}^{m} (1 - E(P_j))} \tag{26}$$

According to Equation (26), the weight of each index can be determined adaptively. Besides, $w_j \in [0,1]$ and $\sum w_j = 1$.

Here we give a brief explanation of the proposed EW-LCB criterion. The term $w_1 \hat{y}(x)$ is used for local exploitation, which concerns the optimal value. On the other side, the term $w_2 \hat{s}(x) \exp((-1)^r r)$ focuses on global exploration, which pays more attention to the uncertainty of the Kriging model for the potential global optimal region. If $w_1 \ll w_2 \exp((-1)^r r)$, it means the algorithm focuses more on global exploration. While $w_1 \gg w_2 \exp((-1)^r r)$ means the algorithm focuses more on local exploitation. The factor $\exp((-1)^r r)$ serves as the catalyst to help the optimization process out of a local optimization solution. However, this factor may decrease the convergence speed of the proposed algorithm because the weight of the exploration will dominate $EWLCB(x)$ when the current optimization solution is repeated too many times. Finally, the point with the minimum value of $EWLCB(x)$ is selected as the new update point.

In this illustrated example, the weight parameters are $w_1 = 0.4961, w_2 \exp((-1)^r r) = 0.5039$ in the first iteration. It is shown that the algorithm focuses more on global exploration than local exploitation. Therefore, another sample point $x = 0.4432$ is added, and the corresponding Kriging model is refreshed, which is shown in Figure 3.

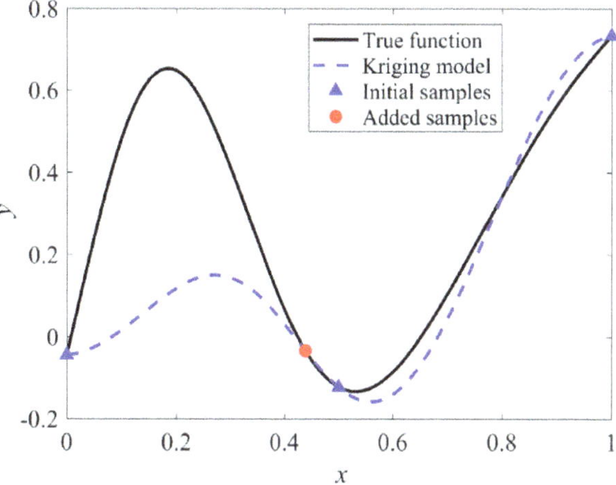

Figure 3. The first iteration of the proposed approach with the illustrated function.

3.5. Step 6: Output the Optimal Solution

Once the terminal conditions are achieved, the optimal solution will be the output. As shown in Figure 4, the optimal solution $x = 0.5312$ is obtained, which equals the global minimum.

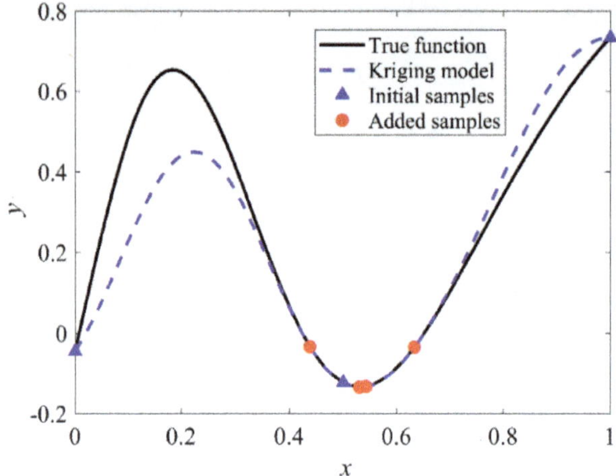

Figure 4. The final result of the proposed approach on the illustrated function.

As shown in Figure 4, the trend of the actual function is recognized by the proposed approach and the optimal value can be obtained although the global accuracy of the Kriging model is not at a high level.

For comparison, four AKBDO methods, Expected improvement infill criterion (EI) [21], weighted expected improvement infill criterion (WEI) [28], Lower confidence bounding infill criterion (LCB) [22], and Parameterized lower confidence bounding (PLCB)[32], were tested on this case. To avoid the randomness of the LHS and GA, all the methods were repeated 100 times. The statistical results including the mean value and standard deviations of the function calls are summarized in Table 2.

Table 2. Comparison results of different approaches of the illustrated example.

Methods	EI	WEI	LCB	PLCB	EW-LCB
Mean Value	7.64	7.86	7.98	7.66	**6.48**
Standard deviations	1.352	1.498	1.301	1.780	**0.505**

As listed in Table 1, the average number of function calls of the proposed approach is less than those of the four AKBDO methods, indicating that the proposed EW-LCB approach performs better than the four AKBDO methods concerning efficiency. Besides, the standard deviation of the proposed EW-LCB approach is the smallest among all the methods, which means that the proposed approach has the best robustness among all the compared methods in this demonstration case.

4. Tested Cases

4.1. Numerical Examples

In this subsection, ten widely used benchmark problems from Ref. [33,50–52] are used to illustrate the effectiveness of the proposed EW-LCB method. Meanwhile, four famous AKBDO approaches, EI, WEI, LCB, and PLCB, are tested to compare with the EW-LCB method. As the optimal solutions for all benchmark problems can be obtained, the terminal condition is defined such that the relative error between the optimal solution obtained from the Kriging model and the true one is within 0.002. Therefore, the number of iterations is regarded as the merit of effectiveness. Considering the randomness of the results, many AKBDO approaches repeat their algorithms dozens of times and provide statistical results [53,54]. Furthermore, some approaches use the deterministic sampling and optimization algorithms such as Hammersley and deterministic PSO [55–57] to avoid the randomness. In this work, each method ran 100 times with different initial samples and their statistical results are recorded in Table 3.

Table 3. The comparison of statistical optimization results.

Functions	Items	EI	WEI	LCB	PLCB	EW-LCB
PK	FE_{mean}	29.82/3	30.13/4	29.68/2	31.99/5	**26.97/1**
	FE_{std}	2.435/2	2.884/3	**2.068/1**	5.107/4	5.34/5
BA	FE_{mean}	33.23/3	32.15/2	33.88/4	34.34/5	**26.33/1**
	FE_{std}	3.989/3	3.777/2	5.664/4	6.144/5	**2.78/1**
SA	FE_{mean}	32.12/3	36.15/5	34.88/4	31.23/2	**27.92/1**
	FE_{std}	4.674/3	4.865/4	2.813/2	5.241/5	**2.722/1**
SC	FE_{mean}	39.30/4	40.66/5	39.20/3	36.41/2	**33.42/1**
	FE_{std}	3.965/3	**3.634/1**	3.785/2	5.292/4	5.320/5
HM	FE_{mean}	45.76/4	46.22/5	44.12/3	41.34/2	**35.22/1**
	FE_{std}	3.456/3	2.973/2	**1.894/1**	5.157/5	3.49/4
GP	FE_{mean}	117.66/5	115.67/4	105.27 /3	97.77/2	**89.27/1**
	FE_{std}	19.11/5	11.81/2	17.55/4	14.44/3	**9.99/1**
GF	FE_{mean}	Failed/5	Failed/5	Failed/5	140.42/2	**116.67/1**
	FE_{std}	Failed/5	Failed/5	Failed/5	75.66/2	**30.63/1**
L3	FE_{mean}	300.4/3	534.6/4	540.4/5	**167.1/1**	199.2/2
	FE_{std}	119.1/3	147.0/4	159.4/5	**55.21/1**	88.89/2
H3	FE_{mean}	37.50/3	37.62/2	38.20/4	39.34/5	**36.58/1**
	FE_{std}	3.50/4	3.46/3	3.29/2	3.64/5	**2.56/1**
H6	FE_{mean}	107.03/4	105.13/3	103.67/2	114.1/5	**101.16/1**
	FE_{std}	44.50/2	50.39/5	48.26/3	49.30/4	**43.43/1**

The expressions of benchmark problems are listed as,

- Peaks function (PK)

$$f(x) = 3(1-x_1)^2 e^{-x_1^2-(x_2+1)^2} - 10(\frac{x_1}{5} - x_1^3 - x_2^5)e^{-x_1^2-x_2^2} - \frac{1}{3}e^{-x_2^2-(x_1+1)^2}, x_{1,2} \in [-3,3] \tag{27}$$

- Banana function (BA)

$$f(x) = 100(x_1^2 - x_2)^2 + (1 - x_1)^2, x_{1,2} \in [-2,2] \tag{28}$$

- Sasena function (SA)

$$f(x) = 2 + 0.01(x_2 - x_1^2)^2 + (1 - x_1)^2 + 2(2 - x_2)^2 + 7sin(0.5x_1)sin(0.7x_1x_2), x_{1,2} \in [0,5] \tag{29}$$

- Six-hump camp-back function (SC)

$$f(x) = (4 - 2.1x_1^2 + \frac{x_1^4}{3})x_1^2 + x_1x_2 + (-4 + 4x_2^2)x_2^2, x_{1,2} \in [-2,2] \tag{30}$$

- Himmelblau function (HM)

$$f(x) = (x_1{}^2 + x_2 - 11)^2 + (x_2{}^2 + x_1 - 7)^2, x_{1,2} \in [-10, 10] \tag{31}$$

- Goldstein–Price function (GP)

$$\begin{aligned} f(x) = \quad & [1 + (x_1 + x_2 + 1)^2(19 - 14x_1 + 3x_1{}^2 - 14x_2 + 6x_1x_2 + 3x_2{}^2)] \\ & \times [30 + (2x_1 - 3x_2)^2(18 - 32x_1 + 12x_1{}^2 + 48x_2 - 36x_1x2 + 27x_2{}^2)], x_{1,2} \in [-2, 2] \end{aligned} \tag{32}$$

- Generalized polynomial function (GF)

$$\begin{aligned} & f(x) = u_1^2 + u_2^2 + u_3^2 \\ & u_i = c_i - x_1(1 - x_2^i), i = 1, 2, 3 \\ & c_1 = 1.5, c_2 = 2.25, c_3 = 2.625, x_{1,2} \in [-5, 5] \end{aligned} \tag{33}$$

- Levy 3 function (L3)

$$\begin{aligned} & f(x) = \sin^2(\pi\omega_1) + \sum_{i=1}^{2}(\omega_i - 1)^2[1 + 10\sin^2(\pi\omega_i + 1)] + (\omega_3 - 1)^2[1 + \sin^2(2\pi\omega_3)] \\ & \omega_i = 1 + \tfrac{x_i - 1}{4}, i = 1, 2, 3, x_i \in [-10, 10] \end{aligned} \tag{34}$$

- Hartmann 3 function (H3)

$$f(x) = -\sum_{i=1}^{4} \alpha_i \exp\left[-\sum_{j=1}^{3} \beta_{ij}(x_j - p_{ij})^2 \right] \tag{35}$$

$$0 \le x_1, x_2, x_3 \le 1$$

where

$$\alpha = \begin{bmatrix} 1 \\ 1.2 \\ 3 \\ 3.2 \end{bmatrix} \beta = \begin{bmatrix} 3.0 & 10 & 30 \\ 0.1 & 10 & 35 \\ 3.0 & 10 & 30 \\ 0.1 & 10 & 35 \end{bmatrix} P = \begin{bmatrix} 0.3689 & 0.1170 & 0.2673 \\ 0.4699 & 0.4387 & 0.7470 \\ 0.1091 & 0.8732 & 0.5547 \\ 0.0381 & 0.5743 & 0.8828 \end{bmatrix} \tag{36}$$

- Leon (LE)

$$f(x) = 100(x_2 - x_1^3)^2 + (x_1 - 1)^2; x_1, x_2 \in [-10, 10] \tag{37}$$

The statistical results of 100 times for five AKBDO approaches are summarized in Table 3. In Table 3, the FE_{mean} represents the mean of iteration times illustrating the efficiency of the method, while FE_{std} denotes the variance of function evaluations, which can reflect the robustness of each method [58]. In Table 3, the numbers after the mean or standard deviation are the rank of the compared method for each numerical case. For example, 26.97/1 means the mean value is 26.97 while the method ranks first. The numbers marked in bold represent the first rank among the five AKBDO approaches. It can be inferred that the EW-LCB ranks first in most of the test problems, which indicates that the proposed EW-LCB outperforms the other compared approaches considering effectiveness. To further demonstrate the robustness of the proposed approach, Figure 5 plots the box plot of FE_{mean} of all 100 runs.

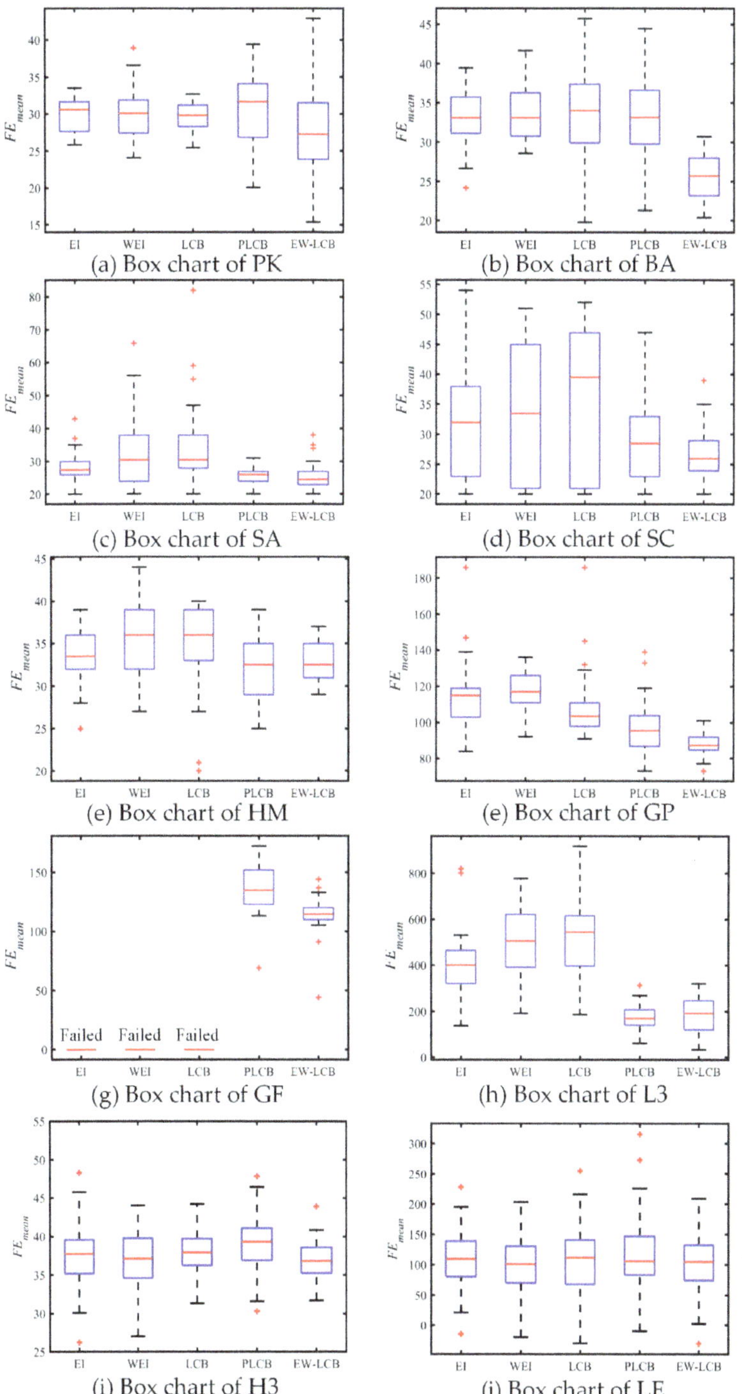

Figure 5. The box plot of the FEmean with different methods.

Table 4 shows the average ranking of the performance of five AKBDO for all the tests. The average ranking of EW-LCB is the best among the five approaches. It is then followed with PLCB, LCB, EI, and WEI. When it comes to the robustness of the compared approaches, the proposed EW- LCB performs better than the PLCB, LCB, and WEI methods, while it is slightly inferior to the EI method. To evaluate whether the differences between the proposed EW-LCB method and the other four approaches are significant or not, the p values over multiple test cases are obtained by using the Bergmann–Hommel procedure [59]. The statistic test results are listed in Table 5. As shown in Table 5, all the p_i-values are less than 0.05, indicating that there are significant differences in the efficiency performance between the proposed EW-LCB and the other four approaches.

Table 4. Average ranking results for all AKBDO approaches considering all numerical cases.

	Metrics	EI	WEI	LCB	PLCB	EW-LCB
Average rank	FEmean	3.70	4.00	3.22	3.10	1.10
	FEstd	2.50	3.40	2.90	3.70	2.30

Table 5. The p-values obtained in the numerical examples by the difference significance test.

i	Hypothesis	p-Values
1	EI vs. EW-LCB	0.0028
2	WEI vs. EW-LCB	0.0001
3	LCB vs. EW-LCB	0.0016
4	PLCB vs. EW-LCB	0.0056

To demonstrate the influences of initial sample sizes, the other two initial sample sizes were studied. The initial sample points are all generated by the LHD method, and function SA and L3 were selected as function SA needs a small sample size, while function L3 needs a large sample size. Table 6 shows the results of comparisons. The numbers after '/' represent the efficiency metric ranking of each method. It is easy to see that the initial sample sizes have a great influence on the function of SA. For function SA, EW-LCB always ranks first, while the ranks of EI, WEI, LCB, PLCB change a lot. For instance, PLCB ranks second when the initial sample sizes are $5d$ and $15d$, while when it comes to the size of $10d$, the rank changes to fourth. As for function L3, the numbers of the sample size of EI, WEI, LCB change little, while the numbers of the sample size of PLCB and EW- LCB change significantly when the initial sample size transforms from $5d$ to $10d$. This represents that PLCB and EW-LCB may perform well with a small sample size in the case of a quite complex function while EI, WEI, and LCB only represent this feature when the function is simple. This is attributed to the objective weighting factors in PLCB and EW-LCB, which are able to allocate factors to balance global exploration and local exploitation. In summary, the EW-LCB method shows the greater ability in balancing between global exploration and local exploitation compared to the other four AKBDO methods.

Table 6. Results of different initial sample points for functions SA and GF.

Functions	Initial Sample Size	EI	WEI	LCB	PLCB	EW-LCB
SA	$n = 5d$	27.55/4	28.98/5	27.02/3	24.12/2	**23.34/1**
	$n = 10d$	32.12/3	36.15/5	34.88/4	31.23/2	**27.92/1**
	$n = 15d$	41.67/4	43.98/5	41.12/3	40.20/2	**38.03/1**
L3	$n = 5d$	393.6/3	524.5/5	513.4/4	**142.5/1**	173.2/2
	$n = 10d$	300.4/3	534.6/4	540.4/5	**167.1/1**	199.2/2
	$n = 15d$	403.4/3	525.4/4	536.5/5	**166.6/1**	206.1/2

4.2. Engineering Application

In this section, an underwater vehicle base design problem is utilized to verify the effectiveness of the proposed method. The base is a braced structure for vibration devices in the hull of an underwater vehicle. The main capability of the base provides a platform for the installation of some imported vibration equipment and avoids the vibration transmitting to the hull directly. Meanwhile, the mechanical impedance of the base has a determination effect in reducing the level of noise. Specifically, the mechanical impedance is expected at a high level under all computational frequencies. The structural profile of the base adjoined to the hull of the underwater vehicle is depicted in Figure 6. The fixed structural and material parameters of the cylindrical shell and the base are listed in Table 7.

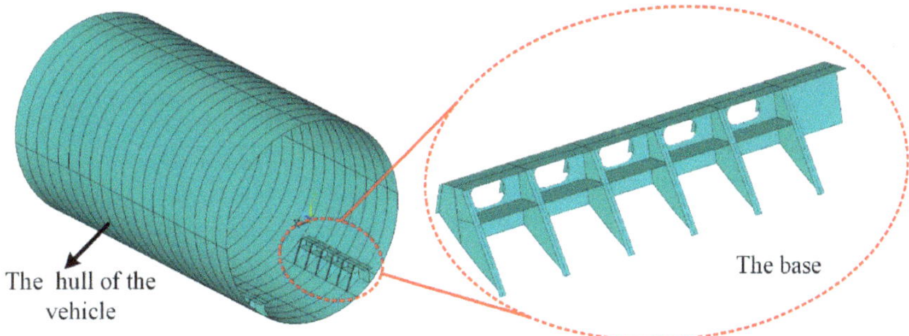

Figure 6. The structural profile of the base.

Table 7. The values of the fixed structural and material parameters.

Fixed Parameters	Values
Elastic modulus E	2.09×10^5 MPa
Density ρ	7850 kg/m^3
Poisson's ratio μ	0.3
The length of the Hull L	12,000 mm
The radius of the Hull R	3300 mm
Rib space l	600 mm
Size of the ribs'	$14 \times 224 / 26 \times 80$ mm
The radius of the base web opening r	75 mm
Width of the base web opening d	210 mm

In this work, the objective is to maximize the minimum difference of the impedance between the scheme in design and the required impedance value under the same frequency. Simultaneously, the weight of the optimized scheme should be less than that of the allowable one. Therefore, the optimization problem can be described as,

$$
\begin{aligned}
&find\ x = [x_1, \ldots, x_6] \\
&\max f(x) = \min\{I(x,\omega_i) - I(x_0, \omega_i); i = 1, 2, \ldots, k\} \\
&s.t.\ \ g(x) = W(x) - W(x_0) < 0
\end{aligned}
\tag{38}
$$

where x represents the vector of the design variables, which is a six-dimensional vector. ω_i is the ith computational frequency. In detail, the design variables of this problem are the thickness of the panels of the base. Figure 7 plots the schematic diagram and Table 8 lists the meanings and value space of the design variables, respectively. $I(x,\omega_i)$ and $W(x)$ represent the impedance value under a specific computational frequency and the weight of the base, respectively. $I(x, \omega_i)$ and $W(x_0)$ are the required impedance value at the ith computational frequency and the allowable base weight, respectively.

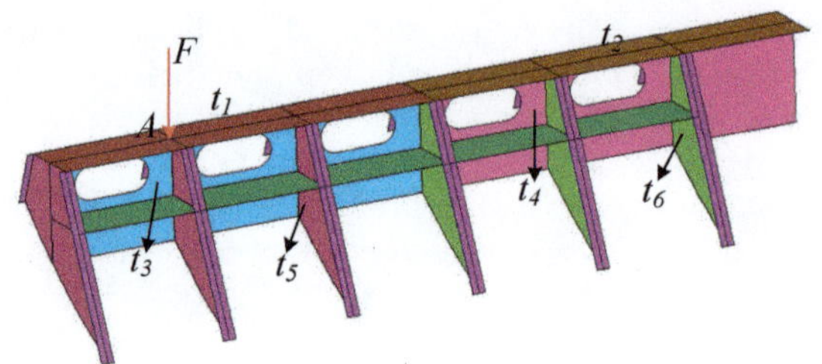

Figure 7. The design variables and loads of the underwater vehicle base.

Table 8. The meaning and ranges of the design variables.

	Design Variables	**Ranges**
Former half	The thickness of the base panel t_1	40–90 mm
	The thickness of the base web t_3	10–60 mm
	The thickness of the base bracket t_5	12–40 mm
Remaining half	The thickness of the base panel t_2	40–90 mm
	The thickness of the base web t_4	10–60 mm
	The thickness of the base bracket t_6	12–40 mm

Generally, the responses of the impedance curve are obtained through the finite element simulation software ANSYS 18.1. The computational platform is with a 4.01 GHz Intel(R) Eight-Core (TM) i9-9900ks Processor and 64 GB RAM. In this simulation, the boundary condition is fixed for all the translation degrees at both sides of the shell. The loading is a unit vertical force at point A as depicted in Figure 7. The ribs are simulated by the Beam 188 element and the rest of the model is simulated by the Shell 181 element. The number of elements has to be more than 34,000 to get a desirable accuracy of the impedance value, as shown in Figure 8. Then, the frequency step is set to be 2.5 Hz and the computational frequency ranges from 0 to 350 Hz. To improve the efficiency of the optimization, minimal convex polygon technology is adopted to pre-process the impedance curve. In that case, the global feature of the curve and the minimum impedance value of the impedance curve are preserved. However, these complex and multimodal features, which may disturb the convergence speed of the optimization process, are filtered. Figure 9 illustrates the impedance curves before and after pre-processing on the scheme $x = [60, 60, 30, 30, 24, 24]$.

Figure 8. The mesh model of the base with the shell of the underwater vehicle.

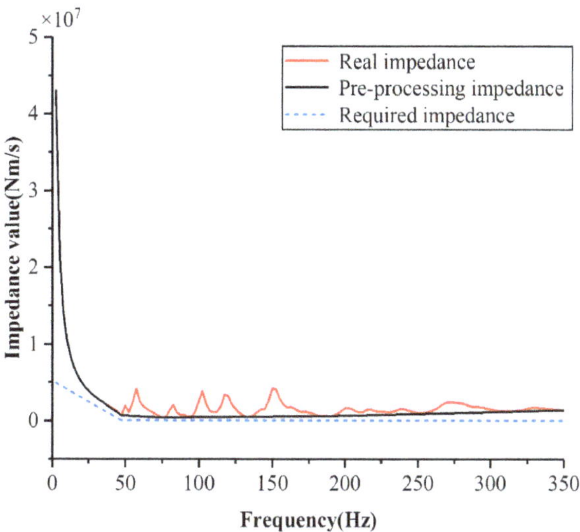

Figure 9. The impedance curves before and after pre-processing.

As shown in Figure 9, the red line denotes the real impedance curve, which fluctuates significantly under different frequencies. The black line is the impedance curve after the pre- processing operation, which is smooth and the minimum value of the original curve remains the same. The blue line is the required impedance curve under different frequencies. Moreover, the allowance weight for this case is 3.027t and the maximum iteration of this case is set to be 50. To eliminate the randomness of the initial samples and the genetic algorithm, all methods use the same 60 initial LHS samples and the same setups of the genetic algorithm. Moreover, all the methods are repeated 20 times to avoid randomness occurring even though the setups are the same. The statistical optimization results of this problem with all compared methods of under 20 runs are summarized in Table 9. Furthermore, the detailed design variables, optimal, and weights of all runs are listed in Table A1 in the Appendix A.

Table 9. The statistical optimization results of the engineering case with different methods.

Methods	$f(x)$ ($\times 10^5$ Ns/m)			
	Max	Mean	Std	Succeeded
EI	4.031	3.998	0.03060	20/20
WEI	4.042	3.983	0.04316	20/20
LCB	3.857	3.733	0.08629	5/20
PLCB	3.892	3.723	0.12230	11/20
EW-LCB	**4.062**	**4.027**	**0.01953**	**20/20**

As illustrated in Table 9, the best value of the proposed method is $4.062 \times 10^5 Nm/s$, which is the maximum optimal value among all methods. Moreover, the proposed method has the maximum mean value on the objective among all the listed methods. It indicates the effectiveness of the proposed approach. Regarding robustness, the proposed method also performs best among all these methods because the proposed method obtains the minimum standard deviation. It is worth mentioning that the LCB and PLCB methods obtain some infeasible solutions. In detail, there are 15 and 9 runs that have failed for the LCB and PLCB methods respectively. The results show that the proposed method is a stable and effective method to solve this engineering optimization problem. Figure 10 shows the impedance curves of the optimal scheme of the proposed approach and the original scheme. As illustrated in Figure 10, the impedance curve of the optimal scheme is better than that of the original scheme because the impedance curve has larger impedance values in the frequencies which are critical to the performance of the base as shown in the sub-figure of Figure 9. On the other side, in the frequencies which are not critical to the performance of the base, the optimal scheme has smaller values than those of the original scheme. Therefore, the proposed approach is an effective method for this engineering case.

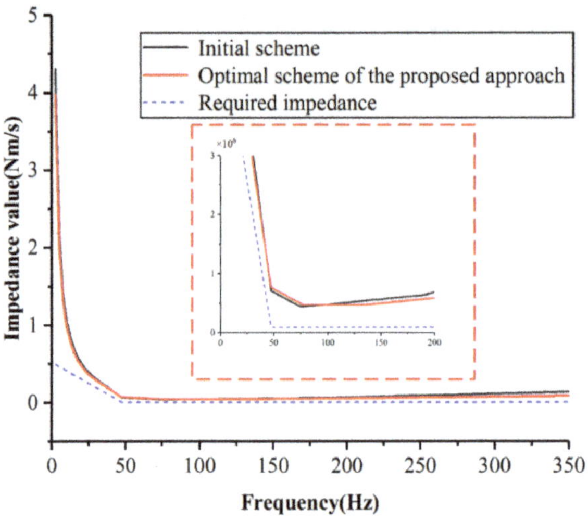

Figure 10. The impedance curve of the optimization solution.

5. Conclusions

To balance exploration and exploitation during the sequential process of the AKBDO, an EW-LCB approach was developed in this work to obtain an optimal solution with less computational resources. In the proposed EW-LCB approach, entropy theory was adopted to quantify the degree of variation of the predicted value and variance of the Kriging model, respectively. Then, the weights were assigned to the LCB function automatically according to the relative values of the entropy theory. Meanwhile, an index factor was defined, which changed with iterations of the appearance of the current optimum, to avoid the sequential process being lost in the local optimum. The updated point was generated by minimizing the EW-LCB function, and the Kriging model updated sequentially. To test the performance of the proposed EW-LCB methods, four typical AKBDO methods including EI, WEI, LCB, and PLCB were adopted for comparison on ten widely used benchmark numerical functions and an engineering case. Results show that the proposed EW-LCB approach can obtain the optimum with the desired accuracy using less computational burden. Moreover, the proposed method has competitive robustness compared with state-of-the-art methods.

It is of note that the proposed method can handle constrained optimization problems by transferring the constrained optimization to the unconstrained one using the penalty methods. In practical engineering cases, simulation models with different fidelities always are available, as part of our future work, the developed EW-LCB method will be extended to the multi-fidelity scenario.

Author Contributions: Conceptualization, J.Q. and J.Y.; methodology, J.Q.; software, J.Q., validation, J.Y. and J.Q.; formal analysis, J.Q.; investigation, J.Q.; resources, J.Q.; data curation, J.Q.; writing—original draft preparation, J.Q.; writing—review and editing, J.Q., J.Y., J.Z., Y.C., and J.L.; visualization, J.Q.; supervision, J.Z., Y.C., and J.L.; funding acquisition, Y.C., and J.L. All authors have read and agreed to the published version of the manuscript.

Funding: This work was supported by the Research Funds of the Maritime Defense Technologies Innovation.

Conflicts of Interest: The authors declare no conflict of interest.

Appendix A

Table A1. Detailed optimization results of the engineering case with different methods.

Methods	Variables (Rounded)						$f(x)$ ($\times 10^5$ Ns/m)	Weight (t) Allowance 3.0327
	t_1 mm	t_3 mm	t_5 mm	t_2 mm	t_4 mm	t_6 mm		
EI	88	59	12	40	16	12	3.945	3.0058
	90	60	12	40	13	13	3.991	3.0151
	90	60	12	40	14	12	4.020	3.0136
	90	59	12	40	15	12	4.031	3.0125
	83	60	12	40	20	12	4.004	3.0158
	88	59	12	40	16	12	3.945	3.0058
	90	60	12	40	13	13	3.991	3.0151
	90	60	12	40	14	12	4.020	3.0136
	90	59	12	40	15	12	4.031	3.0125
	83	60	12	40	20	12	4.004	3.0158
	88	59	12	40	16	12	3.945	3.0058
	90	60	12	40	13	13	3.991	3.0151
	90	60	12	40	14	12	4.020	3.0136
	90	59	12	40	15	12	4.031	3.0125
	83	60	12	40	20	12	4.004	3.0158
	88	59	12	40	16	12	3.945	3.0058
	90	60	12	40	13	13	3.991	3.0151
	90	60	12	40	14	12	4.020	3.0136
	90	59	12	40	15	12	4.031	3.0125
	83	60	12	40	20	12	4.004	3.0158
WEI	88	59	12	40	16	13	3.919	3.0172
	89	60	12	40	14	12	4.042	3.0047
	88	60	12	40	16	12	4.013	3.0172
	83	60	13	40	20	12	3.974	3.0266
	82	60	13	40	21	12	3.964	3.0281
	88	59	12	40	16	13	3.919	3.0172
	89	60	12	40	14	12	4.042	3.0047
	88	60	12	40	16	12	4.013	3.0172
	82	60	13	40	21	12	3.964	3.0281
	83	60	13	40	20	12	3.974	3.0266
	88	59	12	40	16	13	3.919	3.0172
	89	60	12	40	14	12	4.042	3.0047
	88	60	12	40	16	12	4.013	3.0172
	83	60	13	40	20	12	3.974	3.0266
	82	60	13	40	21	12	3.964	3.0281
	88	59	12	40	16	13	3.919	3.0172
	89	60	12	40	14	12	4.042	3.0047
	88	60	12	40	16	12	4.013	3.0172
	83	60	13	40	20	12	3.974	3.0266
	82	60	13	40	21	12	3.964	3.0281
LCB	81	58	16	42	18	12	3.808	3.0203
	78	58	20	40	18	14	3.857	3.0457
	75	52	17	46	17	21	3.676	3.0409
	69	51	25	43	20	19	3.688	3.0617
	76	50	21	44	15	22	3.647	3.0499
	81	58	16	42	18	12	3.808	3.0203
	78	58	20	40	18	14	3.857	3.0457
	75	52	17	46	17	21	3.676	3.0409
	69	51	25	43	20	19	3.688	3.0617
	76	50	21	44	15	22	3.647	3.0499
	81	58	16	42	18	12	3.808	3.0203
	78	58	20	40	18	14	3.857	3.0457
	75	52	17	46	17	21	3.676	3.0409
	69	51	25	43	20	19	3.688	3.0617
	76	50	21	44	15	22	3.647	3.0499
	81	58	16	42	18	12	3.808	3.0203
	78	58	20	40	18	14	3.857	3.0457
	75	52	17	46	17	21	3.676	3.0409
	78	47	21	47	14	20	3.651	3.0303
	70	51	25	49	12	21	3.632	3.0461

Table A1. *Cont.*

Methods	Variables (Rounded)						$f(x)$ ($\times 10^5$ Ns/m)	Weight (t) Allowance 3.0327
	t_1 mm	t_3 mm	t_5 mm	t_2 mm	t_4 mm	t_6 mm		
	78	55	12	40	11	12	3.892	2.8243
	84	47	17	46	12	21	3.769	3.0154
	84	47	24	42	12	19	3.748	3.0331
	45	51	33	46	35	17	3.533	3.1076
	84	38	28	44	15	19	3.684	3.0270
	78	55	12	40	11	12	3.892	2.8243
	84	47	17	46	12	21	3.769	3.0154
	84	47	24	42	12	19	3.748	3.0331
	45	51	33	46	35	17	3.533	3.1076
PLCB	84	38	28	44	15	19	3.684	3.0270
	78	55	12	40	11	12	3.892	2.8243
	84	47	17	46	12	21	3.769	3.0154
	84	47	24	42	12	19	3.748	3.0331
	45	51	33	46	35	17	3.533	3.1076
	84	38	28	44	15	19	3.684	3.0270
	78	55	12	40	11	12	3.892	2.8243
	84	47	17	46	12	21	3.769	3.0154
	84	47	24	42	12	19	3.748	3.0331
	67	59	39	40	11	12	3.645	3.0369
	45	51	35	47	33	16	3.524	3.1050
	88	60	12	40	15	12	4.015	3.0065
	87	60	12	40	17	12	4.017	3.0187
	87	60	12	40	17	12	4.016	3.0187
	85	60	12	40	18	12	4.016	3.0118
	84	60	12	40	18	12	4.015	3.0034
	88	60	12	40	15	12	4.015	3.0065
	87	60	12	40	17	12	4.017	3.0187
	87	60	12	40	17	12	4.016	3.0187
	85	60	12	40	18	12	4.016	3.0118
EW-LCB	84	60	12	40	18	12	4.015	3.0034
	88	60	12	40	15	12	4.015	3.0065
	87	60	12	40	17	12	4.017	3.0187
	87	60	12	40	17	12	4.016	3.0187
	85	60	12	40	18	12	4.016	3.0118
	84	60	12	40	18	12	4.015	3.0034
	89	59	12	40	16	12	4.061	3.0143
	89	58	12	40	16	12	4.056	3.0026
	88	60	12	40	16	12	4.062	3.0172
	89	60	12	40	14	12	4.060	3.0047
	89	60	12	40	15	12	4.059	3.0155

Note: the weights which are larger than the allowable ones are marked by red.

References

1. Han, Z.; Xu, C.; Zhang, L.; Zhang, Y.; Zhang, K.; Song, W. Efficient aerodynamic shape optimization using variable-fidelity surrogate models and multilevel computational grids. *Chin. J. Aeronaut.* **2020**, *33*, 31–47. [CrossRef]
2. Zhou, Q.; Wu, J.; Xue, T.; Jin, P. A two-stage adaptive multi-fidelity surrogate model-assisted multi-objective genetic algorithm for computationally expensive problems. *Eng. Comput.* **2019**, 1–17. [CrossRef]
3. Velayutham, K.; Venkadeshwaran, K.; Selvakumar, G. Process Parameter Optimization of Laser Forming Based on FEM-RSM-GA Integration Technique. *Appl. Mech. Mater.* **2016**, *852*, 236–240. [CrossRef]
4. Gutmann, H.M. A radial basis function method for global optimization. *J. Glob. Optim.* **2001**, *19*, 201–227. [CrossRef]
5. Zhou, Q.; Rong, Y.; Shao, X.; Jiang, P.; Gao, Z.; Cao, L. Optimization of laser brazing onto galvanized steel based on ensemble of metamodels. *J. Intell. Manuf.* **2018**, *29*, 1417–1431. [CrossRef]
6. Sacks, J.; Welch, W.J.; Mitchell, T.J.; Wynn, H.P. Design and analysis of computer experiments. *Stat. Sci.* **1989**, *4*, 409–423. [CrossRef]

7. Zhou, Q.; Wang, Y.; Choi, S.-K.; Jiang, P.; Shao, X.; Hu, J.; Shu, L. A robust optimization approach based on multi-fidelity metamodel. *Struct. Multidiscip. Optim.* **2018**, *57*, 775–797. [CrossRef]

8. Shu, L.; Jiang, P.; Song, X.; Zhou, Q. Novel Approach for Selecting Low-Fidelity Scale Factor in Multifidelity Metamodeling. *AIAA J.* **2019**, *57*, 5320–5330. [CrossRef]

9. Zhou, Q.; Shao, X.; Jiang, P.; Zhou, H.; Shu, L. An adaptive global variable fidelity metamodeling strategy using a support vector regression based scaling function. *Simul. Model. Pract. Theory* **2015**, *59*, 18–35. [CrossRef]

10. Jiang, C.; Cai, X.; Qiu, H.; Gao, L.; Li, P. A two-stage support vector regression assisted sequential sampling approach for global metamodeling. *Struct. Multidiscip. Optim.* **2018**, *58*, 1657–1672. [CrossRef]

11. Qian, J.; Yi, J.; Cheng, Y.; Liu, J.; Zhou, Q. A sequential constraints updating approach for Kriging surrogate model-assisted engineering optimization design problem. *Eng. Comput.* **2019**, *36*, 993–1009. [CrossRef]

12. Shishi, C.; Zhen, J.; Shuxing, Y.; Apley, D.W.; Wei, C. Nonhierarchical multi-model fusion using spatial random processes. *Int. J. Numer. Methods Eng.* **2016**, *106*, 503–526. [CrossRef]

13. Hennig, P.; Schuler, C.J. Entropy search for information-efficient global optimization. *J. Mach. Learn. Res.* **2012**, *13*, 1809–1837.

14. Forrester, A.I.J.; Sóbester, A.; Keane, A.J. *Engineering Design via Surrogate Modelling: A Practical Guide*; John Wiley and Sons: Hoboken, NJ, USA, 2008.

15. Shu, L.; Jiang, P.; Wan, L.; Zhou, Q.; Shao, X.; Zhang, Y. Metamodel-based design optimization employing a novel sequential sampling strategy. *Eng. Comput.* **2017**, *34*, 2547–2564. [CrossRef]

16. Zhou, Q.; Wang, Y.; Choi, S.-K.; Jiang, P.; Shao, X.; Hu, J. A sequential multi-fidelity metamodeling approach for data regression. *Knowl.-Based Syst.* **2017**, *134*, 199–212. [CrossRef]

17. Dong, H.; Song, B.; Dong, Z.; Wang, P. SCGOSR: Surrogate-based constrained global optimization using space reduction. *Appl. Soft Comput.* **2018**, *65*, 462–477. [CrossRef]

18. Dong, H.; Song, B.; Dong, Z.; Wang, P. Multi-start Space Reduction (MSSR) surrogate-based global optimization method. *Struct. Multidiscip. Optim.* **2016**, *54*, 907–926. [CrossRef]

19. Serani, A.; Pellegrini, R.; Wackers, J.; Jeanson, C.-E.; Queutey, P.; Visonneau, M.; Diez, M. Adaptive multi-fidelity sampling for CFD-based optimisation via radial basis function metamodels. *Int. J. Comput. Fluid Dyn.* **2019**, *33*, 237–255. [CrossRef]

20. Pellegrini, R.; Iemma, U.; Leotardi, C.; Campana, E.F.; Diez, M. Multi-fidelity adaptive global metamodel of expensive computer simulations. In Proceedings of the 2016 IEEE Congress on Evolutionary Computation (CEC), Vancouver, BC, Canada, 24–29 July 2016; pp. 4444–4451.

21. Jones, D.; Schonlau, M.; Welch, W. Efficient global optimization of expensive black-box functions. *J. Glob. Optim.* **1998**, *13*, 455–492. [CrossRef]

22. Cox, D.D.; John, S. A statistical method for global optimization. In Proceedings of the 1992 IEEE International Conference on Systems, Man, and Cybernetics, Chicago, IL, USA, 18–21 October 1992; Volume 1242, pp. 1241–1246.

23. Li, G.; Zhang, Q.; Sun, J.; Han, Z. Radial basis function assisted optimization method with batch infill sampling criterion for expensive optimization. In Proceedings of the 2019 IEEE Congress on Evolutionary Computation (CEC), Wellington, New Zealand, 10–13 June 2019; pp. 1664–1671.

24. Diez, M.; Volpi, S.; Serani, A.; Stern, F.; Campana, E.F. Simulation-based design optimization by sequential multi-criterion adaptive sampling and dynamic radial basis functions. In *Advances in Evolutionary and Deterministic Methods for Design, Optimization and Control in Engineering and Sciences*; Springer: Berlin/Heidelberg, Germany, 2019; pp. 213–228.

25. Boukouvala, F.; Ierapetritou, M.G. Derivative-free optimization for expensive constrained problems using a novel expected improvement objective function. *AIChE J.* **2014**, *60*, 2462–2474. [CrossRef]

26. Li, Z.; Wang, X.; Ruan, S.; Li, Z.; Shen, C.; Zeng, Y. A modified hypervolume based expected improvement for multi-objective efficient global optimization method. *Struct. Multidiscip. Optim.* **2018**, *58*, 1961–1979. [CrossRef]

27. Li, L.; Wang, Y.; Trautmann, H.; Jing, N.; Emmerich, M. Multiobjective evolutionary algorithms based on target region preferences. *Swarm Evol. Comput.* **2018**, *40*, 196–215. [CrossRef]

28. Xiao, S.; Rotaru, M.; Sykulski, J.K. Adaptive Weighted Expected Improvement With Rewards Approach in Kriging Assisted Electromagnetic Design. *IEEE Trans. Magn.* **2013**, *49*, 2057–2060. [CrossRef]

29. Zhan, D.; Cheng, Y.; Liu, J. Expected improvement matrix-based infill criteria for expensive multiobjective optimization. *IEEE Trans. Evol. Comput.* **2017**, *21*, 956–975. [CrossRef]

30. Sasena, M.J. Flexibility and Efficiency Enhancements for Constrained Global Design Optimization with Kriging Approximations. Ph.D. Thesis, University of Michigan, Ann Arbor, MI, USA, 2002.

31. Forrester, A.I.J.; Keane, A.J. Recent advances in surrogate-based optimization. *Prog. Aerosp. Sci.* **2009**, *45*, 50–79. [CrossRef]

32. Zheng, J.; Li, Z.; Gao, L.; Jiang, G. A parameterized lower confidence bounding scheme for adaptive metamodel-based design optimization. *Eng. Comput.* **2016**, *33*, 2165–2184. [CrossRef]

33. Cheng, J.; Jiang, P.; Zhou, Q.; Hu, J.; Yu, T.; Shu, L.; Shao, X. A lower confidence bounding approach based on the coefficient of variation for expensive global design optimization. *Eng. Comput.* **2019**, *36*, 1–21. [CrossRef]

34. Srinivas, N.; Krause, A.; Kakade, S.M.; Seeger, M.W. Information-Theoretic Regret Bounds for Gaussian Process Optimization in the Bandit Setting. *IEEE Trans. Inf. Theory* **2012**, *58*, 3250–3265. [CrossRef]

35. Desautels, T.; Krause, A.; Burdick, J.W. Parallelizing exploration-exploitation tradeoffs in Gaussian process bandit optimization. *J. Mach. Learn. Res.* **2014**, *15*, 3873–3923.

36. Yondo, R.; Andrés, E.; Valero, E. A review on design of experiments and surrogate models in aircraft real-time and many-query aerodynamic analyses. *Prog. Aerosp. Sci.* **2018**, *96*, 23–61. [CrossRef]

37. Candelieri, A.; Perego, R.; Archetti, F. Bayesian optimization of pump operations in water distribution systems. *J. Glob. Optim.* **2018**, *71*, 213–235. [CrossRef]

38. Krige, D.G. A statistical approach to some basic mine valuation problems on the witwatersrand. *J. South. Afr. Inst. Min. Metall.* **1952**, *52*, 119–139.

39. Guo, Z.; Song, L.; Park, C.; Li, J.; Haftka, R.T. Analysis of dataset selection for multi-fidelity surrogates for a turbine problem. *Struct. Multidiscip. Optim.* **2018**, *57*, 2127–2142. [CrossRef]

40. Deb, K. An efficient constraint handling method for genetic algorithms. *Comput. Methods Appl. Mech. Eng.* **2000**, *186*, 311–338. [CrossRef]

41. Schutte, J.F.; Reinbolt, J.A.; Fregly, B.J.; Haftka, R.T.; George, A.D. Parallel global optimization with the particle swarm algorithm. *Int. J. Numer. Methods Eng.* **2004**, *61*, 2296–2315. [CrossRef] [PubMed]

42. Volpi, S.; Diez, M.; Gaul, N.J.; Song, H.; Iemma, U.; Choi, K.; Campana, E.F.; Stern, F. Development and validation of a dynamic metamodel based on stochastic radial basis functions and uncertainty quantification. *Struct. Multidiscip. Optim.* **2015**, *51*, 347–368. [CrossRef]

43. Liu, J.; Han, Z.; Song, W. Comparison of infill sampling criteria in kriging-based aerodynamic optimization. In Proceedings of the 28th Congress of the International Council of the Aeronautical Sciences, Brisbane, Australia, 23–28 September 2012; pp. 23–28.

44. Loeppky, J.L.; Sacks, J.; Welch, W.J. Choosing the sample size of a computer experiment: A practical guide. *Technometrics* **2009**, *51*, 366–376. [CrossRef]

45. Jin, R.; Chen, W.; Sudjianto, A. An efficient algorithm for constructing optimal design of computer experiments. *J. Stat. Plan. Inference* **2005**, *134*, 268–287. [CrossRef]

46. Lophaven, S.N.; Nielsen, H.B.; Søndergaard, J. *DACE: A Matlab Kriging Toolbox, Version 2.0*; Technical Report Informatics and Mathematical Modelling, (IMM-TR)(12); Technical University of Denmark: Lyngby, Denmark, 2002; pp. 1–34.

47. Homaifar, A.; Qi, C.X.; Lai, S.H. Constrained optimization via genetic algorithms. *Simulation* **1994**, *62*, 242–253. [CrossRef]

48. Gadre, S.R. Information entropy and Thomas-Fermi theory. *Phys. Rev. A* **1984**, *30*, 620. [CrossRef]

49. de Boer, P.-T.; Kroese, D.P.; Mannor, S.; Rubinstein, R.Y. A Tutorial on the Cross-Entropy Method. *Ann. Oper. Res.* **2005**, *134*, 19–67. [CrossRef]

50. Jiang, P.; Cheng, J.; Zhou, Q.; Shu, L.; Hu, J. Variable-Fidelity Lower Confidence Bounding Approach for Engineering Optimization Problems with Expensive Simulations. *AIAA J.* **2019**, *57*, 5416–5430. [CrossRef]

51. Zhou, Q.; Shao, X.Y.; Jiang, P.; Gao, Z.M.; Zhou, H.; Shu, L.S. An active learning variable-fidelity metamodelling approach based on ensemble of metamodels and objective-oriented sequential sampling. *J. Eng. Des.* **2016**, *27*, 205–231. [CrossRef]

52. Toal, D.J.J. Some considerations regarding the use of multi-fidelity Kriging in the construction of surrogate models. *Struct. Multidiscip. Optim.* **2015**, *51*, 1223–1245. [CrossRef]

53. Jiao, R.; Zeng, S.; Li, C.; Jiang, Y.; Jin, Y. A complete expected improvement criterion for Gaussian process assisted highly constrained expensive optimization. *Inf. Sci.* **2019**, *471*, 80–96. [CrossRef]

54. Bali, K.K.; Gupta, A.; Ong, Y.S.; Tan, P.S. Cognizant Multitasking in Multiobjective Multifactorial Evolution: MO-MFEA-II. *IEEE Trans. Cybern.* **2020**, 1–13. [CrossRef]
55. Pellegrini, R.; Serani, A.; Liuzzi, G.; Rinaldi, F.; Lucidi, S.; Campana, E.F.; Iemma, U.; Diez, M. Hybrid Global/Local Derivative-Free Multi-objective Optimization via Deterministic Particle Swarm with Local Linesearch. In Proceedings of the Machine Learning, Optimization, and Big Data, Third International Conference, MOD 2017, Cham, Switzerland, 14–17 September 2017; pp. 198–209.
56. Pellegrini, R.; Serani, A.; Leotardi, C.; Iemma, U.; Campana, E.F.; Diez, M. Formulation and parameter selection of multi-objective deterministic particle swarm for simulation-based optimization. *Appl. Soft Comput.* **2017**, *58*, 714–731. [CrossRef]
57. Zhang, T.; Yang, C.; Chen, H.; Sun, L.; Deng, K. Multi-objective optimization operation of the green energy island based on Hammersley sequence sampling. *Energy Convers. Manag.* **2020**, *204*, 112316. [CrossRef]
58. Zheng, J.; Shao, X.; Gao, L.; Jiang, P.; Qiu, H. A prior-knowledge input LSSVR metamodeling method with tuning based on cellular particle swarm optimization for engineering design. *Expert Syst. Appl.* **2014**, *41*, 2111–2125. [CrossRef]
59. Garcia, S.; Herrera, F. An extension on "statistical comparisons of classifiers over multiple data sets" for all pairwise comparisons. *J. Mach. Learn. Res.* **2008**, *9*, 2677–2694.

 *applied sciences* 

Article

Integration of Dimension Reduction and Uncertainty Quantification in Designing Stretchable Strain Gauge Sensor

Sungkun Hwang [1], Recep M. Gorguluarslan [2], Hae-Jin Choi [3],* and Seung-Kyum Choi [1],*

[1] George W. Woodruff School of Mechanical Engineering, Georgia Institute of Technology, Atlanta, GA 30332, USA; shwang71@gatech.edu

[2] Department of Mechanical Engineering, TOBB University of Economics and Technology, 06560 Ankara, Turkey; rgorguluarslan@etu.edu.tr

[3] School of Mechanical Engineering, Chung-Ang University, Seoul 06974, Korea

* Correspondence: hjchoi@cau.ac.kr (H.-J.C.); schoi@me.gatech.edu (S.-K.C.); Tel.: +82-2-820-5787 (H.-J.C.); +1-404-894-9218 (S.-K.C.)

Received: 29 November 2019; Accepted: 13 January 2020; Published: 16 January 2020

Abstract: Interests in strain gauge sensors employing stretchable patch antenna have escalated in the area of structural health monitoring, because the malleable sensor is sensitive to capturing strain variation in any shape of structure. However, owing to the narrow frequency bandwidth of the patch antenna, the operation quality of the strain sensor is not often assured under structural deformation, which creates unpredictable frequency shifts. Geometric properties of the stretchable antenna also severely regulate the performance of the sensor. Especially rugged substrate created by printing procedure and manual fabrication derives multivariate design variables. Such design variables intensify the computational burden and uncertainties that impede reliable analysis of the strain sensor. In this research, therefore, a framework is proposed not only to comprehensively capture the sensor's geometric design variables, but also to effectively reduce the multivariate dimensions. The geometric uncertainties are characterized based on the measurements from real specimens and a Gaussian copula is used to represent them with the correlations. A dimension reduction process with a clear decision criterion by entropy-based correlation coefficient dwindles uncertainties that inhibit precise system reliability assessment. After handling the uncertainties, an artificial neural network-based surrogate model predicts the system responses, and a probabilistic neural network derives a precise estimation of the variability of complicated system behavior. To elicit better performance of the stretchable antenna-based strain sensor, a shape optimization process is then executed by developing an optimal design of the strain sensor, which can resolve the issue of the frequency shift in the narrow bandwidth. Compared with the conventional rigid antenna-based strain sensors, the proposed design brings flexible shape adjustment that enables the resonance frequency to be maintained in reliable frequency bandwidth and antenna performance to be maximized under deformation. Hence, the efficacy of the proposed design framework that employs uncertainty characterization, dimension reduction, and machine learning-based behavior prediction is epitomized by the stretchable antenna-based strain sensor.

Keywords: stretchable antenna-based strain sensor; structural optimization; structural health monitoring; dimension reduction; entropy-based correlation coefficient; multidisciplinary design and analysis; uncertainty-integrated and machine learning-based surrogate modeling

---

1. Introduction

Structural health monitoring (SHM) is implemented to evaluate the physical conditions of structures with consistent surveillance. In accordance with the information garnered by an SHM system, engineers can identify the critical roots in structural damage or deterioration and provide applicable approaches to avoid structural failures. Compared with traditional fixed inspections with interval schedules that demand excessive maintenance [1], the SHM systems can operate for condition-based maintenance and reduce preventive maintenance cost, life cycle costs, and potential catastrophic failure [2]. Therefore, the systems can heighten both the capability and reliability of the monitoring system. Application of the SHM systems has been broadened from heavy mechanical equipment or civil structures to aerospace or bio-mechanical examples [3].

In the SHM systems, a strain is regarded as a vital factor, which should be rigorously examined. By measuring strain of structures, a strain gauge is able to predict certain failure modes such as crack propagation, deformation [4], vibration, or mechanical loading [5]. Between copious types of strain gauges, metallic foil pattern is mainly employed because it identifies electrical resistance connected to strain changes under structural deformation. In spite of the benefits of a foil gauge (e.g., simple circuit structure, lower fabrication cost, and applicability in various examples), its usability has been restrained owing to the reliance on long cables corresponding to the power supply and data transmission. An acceptable solution to the issue is to use sensors of passive wireless strain gauge, which reduces dependency on batteries and has lower installation costs. In recent years, great attention has been paid to passive wireless antennas on account of their compelling suitability to observe strain of structures [6].

An antenna installed on the surface of a deformable structure transfers details about the strain variations. It is approachable to evaluate the stability and physical circumstance of the structure with the transferred data of the strain variations. In the strain gauge markets, interests in micro-strip patch (MSP) antenna have increased because of lower fabrication cost, lighter load, lower profile planar configuration, and capability of multiband operation [7]. Thus, the majority have still exploited a rigid patch antenna to capture strain variation. However, the conventional rigid patch antenna often fails to deliver reliable information owing to two critical drawbacks: mediocre consideration for non-uniform geometric uncertainties [8] and limited bandwidth of resonance frequency (Rf) [9]. To assure the performance of the MSP antenna-based strain sensor, these two drawbacks should be meticulously examined.

The first drawback is about the geometric uncertainties on the surface of the substrate of an antenna shown in Figure 1a. As depicted in Figure 1b, a typical scanning electron microscope (SEM) shows that the substrate surface has certain scratched patterns, which makes the substrate rugged, so there exist obtrusive thickness variations on the substrate. Variation of the substrate roughness is unavoidable because the unstable fabrication process requires manual operation by engineers [10]. The geometric uncertainties due to this surface roughness are more critical for the stretchable antenna-based strain sensors, which broaden the applicability of such sensors for the wearable devices or human body. Compared with conventional rigid antennas, the uncertainties have a much stronger influence on the performance of the antenna along with the non-uniform changes under deformation [11–13]. Therefore, the substrate fabrication has been steadily operated with polydimethylsiloxane (PDMS) or other stretchable materials to be firmly attached to the surface of rough structures [14].

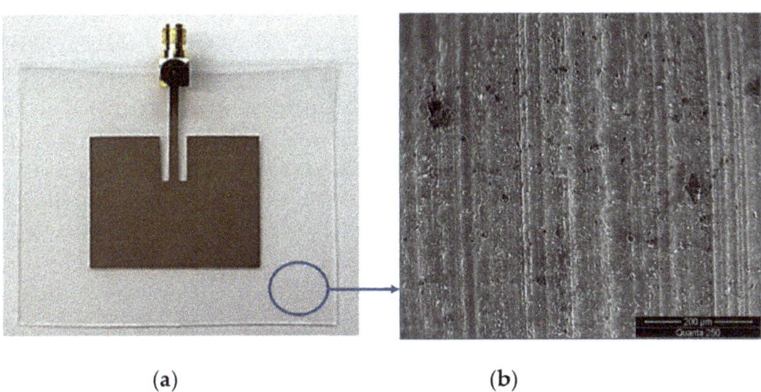

(a)   (b)

Figure 1. Stretchable antenna. (**a**) The antenna (dark color) and the polydimethylsiloxane (PDMS) substrate (light color); (**b**) scratched surface of the substrate observed by FEI Quanta 250 scanning electron microscope (SEM) device.

In the computer-based design and analysis of stretchable antennas for a strain gauge, usually, an assumption of uniform thickness is made to reduce design complexity and computational burden. However, if the randomness corresponding to the thickness is overlooked, reliable response assessment of the MSP antenna-based strain sensor can be impeded. As already proven by various simulations and measurements [15–20], although the analysis of the non-uniform substrate-based antenna is always more complicated than that of constant substrate-based antenna, it involves better reliability predictions as the approach corresponding to the uneven substrate represents the real environment. Thus, it is absolutely imperative to examine the thickness variation to improve the reliability of the antenna-based sensor.

Unfortunately, however, considering the roughness would bring extremely increased computational costs as well as total number of design variables (DVs). It is indispensable to capture and propagate all possible critical variables and the corresponding uncertainties to ensure that the system is reliable, but such an approach would incur high dimensions of DVs, increasing computational complexity [21]. Furthermore, once the DVs are contaminated by the immensely correlated random behavior, the extravagant variables obstruct the reliability of the performance estimation [22,23]. To eliminate these issues, a rigorous design of stretchable electronics based on efficient capturing and modeling of correlated and high dimensional random variables is required [24]. Therefore, this study proposed a dimension reduction (DR) framework to effectively manage uncertainties especially occurring for the substrate thickness. In the framework, a Gaussian copula is utilized [25] to precisely model the intricacy of DVs. The intricacy can be described by a joint distribution that consists of manifold marginal distributions. After modeling the geometric uncertainties, a clear guideline engaging the entropy-based correlation coefficient was exploited to suggest a better decision between feature extraction and selection. Through the guideline, engineers can embrace a competent choice of an efficient DR method based on multivariate data properties. In accordance with the entropy-based correlation coefficient, two DR methods, particularly feature extraction and feature selection, diminish the overabundance of the DVs. Specifically, the performance of two feature extraction methods (principal component analysis (PCA) [26] and auto-encoder (AE) [27]) are evaluated to identify the one that gives the best predictions, while feature selection employs independent features test (IFT) [28]. Details will be addressed in the following sections.

The second drawback is about the narrow bandwidth, which often fails to warrant the operation of the MSP antenna for measuring strain. In general, information transferred by a radio frequency system is only trustworthy within the desired frequency bandwidth. Owing to the unforeseen structural deformation that produces extreme frequency shifts, the stretchable strain sensor cannot perform stable detection when the frequency deviates from an allowable bandwidth. To handle this issue, manifold solutions have been proposed. Specifically, the ultra-wideband (UWB) wireless system is used to facilitate multiple broadband and a high data rate, which enlarges the range for frequency activity [29]. However, the solution is not convincing owing to the complicated design process, power limitation, and standardization. Other approaches engaging optimization techniques also exist, but most have focused on sizing adjustment of the antenna [30]. Even though the sizing optimization can draw simple and efficient antenna update [31], the technique forces the antenna to maintain its original shape, which interrupts a fundamental solution regarding the narrow bandwidth. Furthermore, topology optimization with solid isotropic material with penalization or the level set method has been exploited to address the issue [32], but the efficiency of the suggested design is only identified when it is fabricated by special additive manufacturing process, which intensely increases the fabrication burden. Assorted studies have also suggested unique shape modification such as conical, pentagon, or even fractal geometry [33,34], but such modifications refuse to preserve an initial outline of the MSP antenna that has advantages of cheaper and easier fabrication process and diverse applicability, so their efficiency is only meaningful in a specialized example.

To address these issues of the existing design approaches for the stretchable antenna, a design optimization process that includes a structural shape optimization is proposed in this study. In comparison with the existing approaches, the proposed design process not only inherits the advantages of the MSP antenna, but also derives an innovative antenna shape to enhance functionality in the multi-physics domain. Owing to the mechanical flexibility and electrical radio frequency behavior, the stretchable MSP antenna for strain gauge requires a meticulous multidisciplinary design optimization. Thus, the proposed design maximizes the frequency stability and antenna performance, referring to a return loss under unexpected structural deformation. The optimum design is garnered in accordance with the basic shape of the MSP antenna, so that it can truncate an intense fabrication burden. The design also operates at dual bandwidth to keep the benefit of UWB, but it induces less design intricacy. For the system response prediction, the proposed framework employs the artificial neuron network (ANN)-based surrogate model under geometric uncertainties stemming from the stretchability and fluctuating antenna substrate. Finally, the reliability of the proposed design is evaluated by the probabilistic neuron network (PNN) and Monte Carlo simulation (MCS).

This paper is structured as follows. Section 2 introduces backgrounds of the R_f fluctuation under structural deformation and the DR methods. Section 3 introduces the proposed design framework employing the DR methods and shape optimization to determine the optimal new design of the stretchable antenna-based strain sensor. In Section 4, an example of the MSP antenna employed as a stretchable strain gauge is provided to represent the efficacy and merits of the proposed framework.

2. Micro-Strip Patch Antenna and DR Process

2.1. Micro-Strip Patch Antenna

An MSP antenna is composed of a patch, a feed line, a dielectric substrate, and a ground, as depicted in Figure 2a. Owing to the effect of the inductance and the capacitance, intrinsic resonance frequency (R_f) is produced where inductive and capacitive reactance offset each other, which results in a transfer of signal and power. The R_f fluctuates as the antenna geometry is changed, which directly indicates how much the antenna is distorted. The antenna employed in a strain gauge normally adheres to the fragile surface of a structure, and the antenna's shape will be changed in which the structure is deformed. R_f is then promptly changed in accordance with the structural deformation of the antenna. Therefore, it is ineludible to monitor the R_f variation of the antenna to observe the status of the structured deformities [4]. As shown in Figure 2b, the R_f can be appraised based on the geometric information as follows:

$$R_f = \frac{1}{2(2\Delta L_{le} + L_e)\sqrt{\varepsilon_{re}}}c_l,\tag{1}$$

where ΔL_{le}, L_e, ε_{re}, and C_l indicate variation of the electrical length, the phase length, the effective dielectric constant, and the velocity of the light, respectively. Here, the phase length stands for the geometric dimension of the patch along the direction of antenna's line extension and radiation mode. The effective dielectric constant is rephrased by

$$\varepsilon_{edc} = \frac{\varepsilon_r - 1}{2\sqrt{(1 + 12H/W_e)}} + \frac{\varepsilon_r + 1}{2}, \tag{2}$$

where W_e, H, and ε_r represent the patch's electrical width, constant thickness, and substrate's dielectric constant, respectively. On the basis of Equation (2), the phase length, ΔL_{le}, is denoted by the dimensions:

$$\Delta L_{le} = \frac{(W_e/H + 0.264)(\varepsilon_{re} + 0.3)H}{2.427(W_e/H + 0.813)(\varepsilon_{edc} - 0.258)}. \tag{3}$$

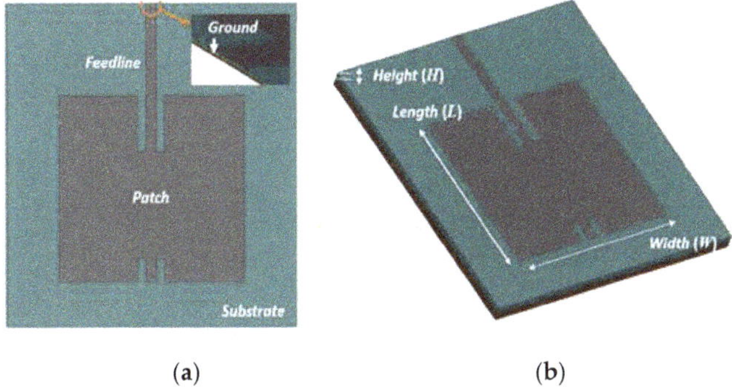

Figure 2. Micro-strip patch (MSP) antenna. (**a**) Composition of the MSP antenna; (**b**) parts of the patch and the substrate.

The R_f fluctuation depends on the path of the tensile strain (ε_L). While an observed R_f that is higher than the initial R_f implies the strain was employed in the electrical width direction (L_e), the value of the R_f should be less under the strain that is applied along electrical length direction (W_e). The frequency oscillation is caused by Poisson's effect by

$$W_e = \left(1 - v_p \varepsilon_L\right) W_o \quad and \quad H = (1 - v_s \varepsilon_L) H_o, \tag{4}$$

where v_p, W_o, v_s, and H_o denote the Poisson's ratios of the patch, the original width, the Poisson's ratios of the substrate, and the thickness, respectively. By a variation of the tensile strain, (W_e/H) is elucidated as independent in the case where the proportion of width to thickness is less than one. Here, the Poisson's ratios, v_p, and v_s, will be the same, leading to the same ε_{re} and ε_r. Hence, the R_f when a strain, ε, is applied along the vertical direction will be denoted by

$$R_f = \frac{c_l}{2\sqrt{\varepsilon_r}} \frac{1}{(1-\varepsilon)(L_e + 2\Delta L_{le})} \approx \frac{R_{of}}{1-\varepsilon} \approx R_{of}(1+\varepsilon), \tag{5}$$

where R_{of} indicates the R_f that is estimated by W_o and H_o. In this study, an example of the tensile strain applied along the width direction is demonstrated. The patch and substrate are able to derive significant effects on the antenna performance. Existing research has suggested ways to intensify the productivity of the antenna by modifying copious antenna components [35,36], but there exists no rigorous deliberation in terms of the variation of the substrates thickness so far [7]. Such deliberation regarding the variations is essential because most fabrication of the substrate is manual. Also, the interests in printable or wearable antenna applications make the deliberation of non-uniform surfaces valuable. Therefore, it is fairly necessary to assume that the substrate thickness varies; however, these additional concerns demand excessive growth in the number of corresponding responses and DVs. When the engineers decide to weight the additional intricacy, it is typical to encounter the curse of dimensionality, which demands the DR methods in the design process.

Copious DVs result in the curse of dimensionality, which triggers unreliable data assessment in classification and function approximation [37]. Furthermore, the multivariate variables are strongly correlated and arouse the redundancy of the data. The higher redundancy imposes negative impacts on the estimation of the system performance because the inter-correlated data is probably required to represent properties already taken into account. To describe the redundancy, Entropy [38] is often exploited. The uncertainties of a certain random variable are measured by Entropy (H) exploiting a probability distribution [39]. Here, x referring to a random variable H is represented by the following:

$$H_X = \sum_{i=1}^{m} p(x_i) \log_2 \frac{1}{p(x_i)}, \tag{6}$$

where $p(x_i)$ means the marginal probability of x. According to the H, the redundancy (R_e) is able to be re-written as follows:

$$R_e = \log_2 N - \sum_{i=1}^{n} p(x_i) \log_2 \frac{1}{p(x_i)}, \tag{7}$$

where $\log_2 N$ indicates the maximum H with the entire samples (N). In order to elevate the assessment accuracy of the system performance, the DR approach is conducted in this research. The DR technique is generally conducted by the main methods: feature extraction (FE) [40] and feature selection (FS) [41]. While FE transforms the initial features located in the higher dimension to lower-dimensional new features, FS takes the most substantial subgroup of the raw features. Details of the DR methods will be explained in the following sections.

2.2. Feature Extraction (FE)

FE builds distinct features in the lower dimension using a conversion of the initial features. This DR method can be interpreted as a mapping, which reduces the R_e of original data. The newly extracted features include most related properties from the initial data. Two major methods, principal component analysis (PCA) [26] and auto-encoder (AE) [42], are explained in this research.

2.2.1. Principal Component Analysis (PCA)

To draw independent and uncorrelated features from heavy correlation, PCA employs orthogonal transformation, which is represented by $Y = PX$. The transformed matrix, Y, is explained by a transformation matrix, P, and the original data set, X. In order to conduct a linear transformation, X and Y will contain numerous examinations and variables, which can be described as m and n, respectively. Specifically, PCA requires a decomposition of eigenvectors [26] to truncate the data dimension. A covariance matrix, C_Y, can be acquired by

$$C_Y = \frac{1}{n-1} PAP^T = \frac{1}{n-1} P(P^T DP)P^T = \frac{1}{n-1} D. \tag{8}$$

In Equation (8), $A = XX^T$, which implies a symmetric condition. A consists of the eigenvectors in the rows of the matrix, P, and is denoted by $P^T DP$. Here, D stands for a diagonal matrix that should be connected to equivalent eigenvectors of matrix A. The property of an orthonormal matrix allows the inverse of the matrix P to be equal to its transpose. Thus, principal components corresponding to X will act as the eigenvectors of $PC_x = (n-1)^{-1} XX^T$, and the diagonal variables of C_Y should be the $X's$ variance.

2.2.2. Auto-Encoder (AE)

While PCA sometimes faces its limitation in non-linear data analysis [26], auto-encoder (AE) is not restricted by the drawback of data linearity. As a special case of the artificial neural network (ANN) [42], AE handles the DR by converting the raw data. AE generally includes three different layers, which are the input, output, and hidden layer, respectively. AE also contains a reconstruction process of the raw data in the output layers. The data will be transformed with a bias and a weight function in the hidden layer. To reduce the gap between the raw and newly recreated data, AE tries to minimize the mean squared error corresponding to the reconstruction [43]. The reconstruction error (r_e) is evaluated by the squared-error cost function,

$$r_e = 0.5 \| \sum_{k=1}^{n} W_k{}^T \left(\sum_{k=1}^{n} W_k x_k + b \right)_k + b^T - X \|^2 , \tag{9}$$

where W, T, x, and b refer to the weights, transpose of the respective vector or matrix, input units, bias, and input vector, respectively. Here, X stands for input vector. A sparsity constraint exploits to minimize the total number of hidden units, so that AE can be conducted as the DR. Owing to the constraint, AE has a bottleneck shape. Some of AE's neurons will be inoperative by applying the sparsity constraint to the hidden units. The hidden layer's average activation can be defined by

$$\Psi_j = \frac{1}{n} \sum_{i=1}^{n} \left[a_j^{(k)} (x^{(i)}) \right], \tag{10}$$

where $a_j^{(k)}$ denotes the activation of hidden layer k and the hidden unit j. To promote the sensitivity accuracy, an extra penalty term (P) is considered.

$$P = \sum_{j=1}^{s} \rho \log \frac{\rho}{\Psi_j} + (1-\rho) \log \frac{1-\rho}{1-\Psi_j}, \tag{11}$$

where s and ρ represent hidden layers and the sparsity parameter, respectively. When the activation value becomes zero, a relation of $\Psi = \rho$ is acquired.

2.3. Feature Selection (FS)

Unlike FE, FS does not take into account the same dimension when the initial data are reconstructed. It just focuses on grabbing the most significant subgroup in the lower dimension. The importance of the elected subset may not be ensured if the subset contains meaningless properties of the initial data. Diverse studies have been achieved using the mutual information method, genetic algorithm, and single variable classifier method [43], but those methods still have handicaps, that is, huge computational complexity and untrustworthy assessment of the data distributions if the behavior of the distributions is not clearly provided [22]. On the other hand, independent features test (IFT), called the simple hypothesis test [27], is able to quickly discard unnecessary features. For this advantage, IFT will be employed in this proposed framework. The target data are assumed as categorical, so IFT allows all features to be assigned to two categories. This approach is conducted by a scoring value of informative features, SVIF.

$$SVIF(F) = \frac{\left| \mu.(A) - \mu.(B) \right|}{\sqrt{\left(\frac{v.(A)}{n_1} + \frac{v.(B)}{n_2} \right)}} > sv \tag{12}$$

From Equation (12), two data sets related to the feature (F)'s values can be described as A and B, where n_1 and n_2 denote the number of the features employed in the categories. With variance, v, and mean, μ, the significance value of data can be calculated. In accordance with a threshold, sv, indicating the significance value to eliminate unimportant features, IFT is able to get the best features. In general, it is suggested that the significance value is equal to or greater than 2 [27].

3. DR-integrated Framework for Optimized Design

On the basis of the developed framework, the DR procedure will be combined with the shape optimization. Even if the framework is expressed with an MSP antenna example, the framework is generalized to be applicable for different design applications, which can be affected by the issue of the curse of the dimensionality. The framework is composed of four steps, as represented in Figure 3. Step 1 shows multivariate input variables that will be randomly created. From Step 2, R_e of the variables will be minimized by the DR, improving computational productivity. In Step 3, the framework integrated with the DR will find the optimal design. Finally, the reliability of the optimized design of an MSP antenna will be assessed.

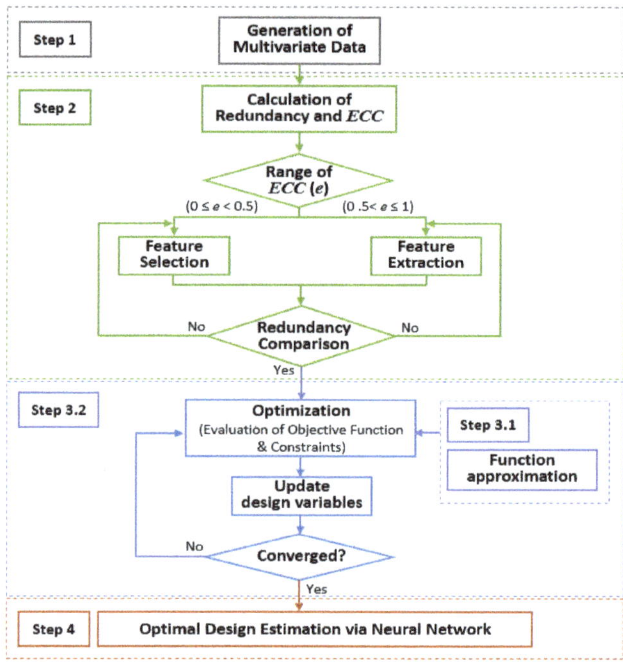

Figure 3. Proposed optimum design framework with the dimension reduction (DR). ECC, entropy-based correlation coefficient.

3.1. Generation of Multivariate Data Random Thickness

The design framework in Figure 3 starts with the generation of random multivariate data including a strong correlation in Step 1. For example, in the design space of the antenna application in this study, the correlation will be considered for the substrate thickness. A Gaussian copula is proposed to be used in this study to meticulously create multi-dimensional random behavior. In comparison with the conventional methods that require a linear correlation coefficient [44], the copula takes two clear benefits: (1) it creates multi-dimensional data, although the DVs contain each distinct marginal distribution; and (2) it analyzes the correlation between the random variables that have non-linear behavior. Hence, in Step 1, the most realistic properties in the multi-dimensional space will be demonstrated. A joint distribution, J_{XY}, is denoted by the two distinct marginal distributions, F_1 and F_2,

$$J_{XY}(x,y) = CP(F_1(x), F_2(y)). \tag{13}$$

Here, CP represents the copula function (CP: $[0,1]^2 \to [0,1]$),

$$CP_\rho(x,y) = \frac{1}{2\pi\sqrt{(1-\rho^2)}} \int_{-\infty}^{\Phi_1^{-1}(x)} \int_{-\infty}^{\Phi_1^{-1}(y)} \exp\left(-\frac{h^2-2\rho hk+k^2}{2(1-\rho^2)}\right) dhdk,$$
$$\Phi(x) = \frac{1}{2\pi} \int_{-\infty}^{x} e\left(-0.5s^2\right) dh, \tag{14}$$

where, ρ, x, and y stand for the linear correlation coefficient, and the marginal distributions of both x and y. Moreover two different copula parameters are represented by h and k, and the standard univariate Gaussian distribution is denoted by Φ.

Therefore, the copula function builds the sample dataset by exploiting the parametric multivariate distribution. The samples can then be generated from the copula and can be utilized for stochastic analysis to model and simulate a complex engineering system.

3.2. Entropy-Based Correlation Coefficient (ECC)

Step 2 of the proposed framework involves the DR process. In the high dimensional system analysis, it is required to make an obvious guideline to decide which DR method (i.e., FE or FS) should be accurately employed. In the framework, an explicit criterion is presented to properly select either FE or FS. The criterion can be exploited in accordance with a status of random variables' correlation. As a conventional method, linear correlation estimation [45] measures the degree of the correlation between the data sets. But the approach cannot be supported once features contain non-Gaussian distributions, non-linear correlation, or correlation coefficient of 0.5. To overcome these challenges, it is inevitable to provide a distinct criterion corresponding to an entropy-based correlation coefficient (ECC) denoted by e [46]. A joint entropy, H_{XY}, is calculated by

$$H_{XY} = \sum_{i=1}^{n}\sum_{j=1}^{n} p_{joint}(x_i)p_{joint}(y_j) \log_2 \frac{1}{p_{joint}(x_i)p_{joint}(y_j)} \tag{15}$$

where $p_{joint}(x_i)$ and $p_{joint}(y_j)$ are the joint probability distributions of x_i and y_j, which are the random variables. As a basic concept of mutual information (MI), entropy can be used to measure dependence, but it does not have a satisfactory scale as the maximum value depends on the size of samples [47]. Hence, the ECC from the normalized MI rescales the MI values to be between 0 and 1. The status of the dependence between the random variables is then calculated by e:

$$e = \sqrt{\frac{H_X + H_Y - H_{XY}}{H_X + H_X}}, \quad 0 \le e \le 1. \tag{16}$$

Here, 1 refers to heavy correlations, and vice versa. On the basis of two different ranges of e, $0 \le e < 0.5$ and $0.5 \le e < 1$, the proposed framework can provide a guidance to use either FE or FS. The first range indicates that the behavior of the variables is independent and uncorrelated. Thus, the FS will minimize the size of the multi-dimensional data. The second range, on the other hand, means that there is a correlation between multivariate data. Hence, FE will be suitable to reduce the redundancy of the data, R_e, originated from the strong correlation between multivariate data. When the value of e in Equation (16) is close to 0.5, it suggests engineers may employ both DR methods. After the reduction proceeds, the comparison of R_e values calculated by Equation (7) is required to confirm whether or not sparse features are sufficiently acquired. If the raw features still include higher R_e compared with the gleaned new features, an additional DR process might be required. Thus, the combination of e with R_e

can be considered as a reliable solution that can draw independent and sparse features by reducing the computational cost.

3.3. Optimization

Once the DR process is conducted in Step 2, an optimization process is then utilized in Step 3 for the optimal design of the application structure or geometry. A function approximation of the actual physical calculations is often required to alleviate the computational burden during the iteration process of the optimization. Therefore, a machine learning-based surrogate modeling technique, namely the artificial neural network (ANN) [27], will be exploited in this research. Compared with the traditional surrogate modeling approaches conducted by linear programming [48], decision trees [49], and discriminant analysis [50], ANN possesses two major benefits: (1) the credibility of ANN is assured in which the decision domain holds complex shapes that are difficult to secure, and (2) ANN is able to be conducted for both classification and function approximation. Essentially, ANN is preferred to be used for the function approximation approach to take the complex contours in the decision domain.

3.4. Predicting Reliability of the Obtained Optimal Design

Once the DR and the optimization are exploited, the reliability of the acquired antenna's optimum design will be checked. This might be a noncompulsory step based on the ascribed design problem. Generally, when a multi-dimensional system is evaluated, it is frequently computationally exorbitant to analyze further response statistics such as reliability of the system. Hence, although ANN is employed for the function approximation during the optimization process in Step 3, in Step 4, another neural network method, called the probabilistic neural network (PNN), is utilized as a classification process is needed for the reliability assessment. As a major advantage, PNN does not require expensive computational costs because of the fast and straightforward training procedure when compared with ANN. The performance of PNN is assured by the Parzen nonparametric estimator and the Bayes decision rule, which lessen the predicted risk of misclassification [51,52].

The following section presents the applicability of the proposed framework by a design example of the stretchable MSP antenna.

4. Stretchable MSP Antenna-Based Strain Sensor

In general, antennas that perform in a single frequency band become a barrier of careful surveillance because of the harmful radiation of wireless components and the interruption of data signal transmission. Hence, the efficacy of the single-band antenna cannot to be ensured. For this reason, a dual-band antenna should be utilized to improve the reliability of the stretchable strain-based MSP antenna. In this example, a dual-band antenna functioning at 2.5 GHz and 5 GHz, usually exploited for wireless fidelity, was regarded. On the basis of composition of the MSP antenna, as shown in Figure 2, vital DVs and the material properties are listed in Table 1. Here, μ and COV stand for mean and coefficient of variation, respectively.

Table 1. Geometric and material properties of the stretchable micro-strip patch (MSP) antenna.

	Substrate (μ, COV)	Patch	Feed Line	Ground	Source (μ, COV)
Thickness (mm)	0.97, 0.05	0.03	0.03	0.05	
Width (mm)	70	47	2.5	70	0.09, 0.1
Length (mm)	70	40	32	70	2.5
Permittivity	2.2	1	1	1	1
Young's Moduli (MPa)	1.32	0.0124	0.0124	1.32	
Conductivity (S/cm)	0	1.51×10^4	1.51×10^4	1.51×10^4	0
Dielectric Loss Tangent	0.0009	0.01	0.01	0.01	0
Magnetic Loss Tangent (kg/m^3)	0	0.001	0.001	0.001	0

To investigate the behavior of stretchable antennas, a multi-physics analysis that includes both the strain estimation under mechanical deformation of the antenna and the R_f estimation by electrical analysis is required. In the previous research [8], such a multidisciplinary analysis was performed for a stretchable antenna geometry and it was shown that the fluctuating thickness of the substrate resulted in a high return loss. This result was because the uncertainties of the substrate thickness led to the escalation of forged feed radiation and surface waves, which limit the range of the bandwidth. Therefore, in this proposed framework, a design optimization process was also utilized to find the optimal contour of the antenna patch to maximize antenna efficiency by reducing the return loss. The existing geometry in the work of [8] was used as the starting geometry in the optimization process; therefore, first, the dimension reduction and prediction using Steps 1 and 2 of the proposed framework were used for the starting geometry.

In the analysis of the antenna, first, a displacement that results in a tensile strain along the width direction, which is denoted as the y direction on Cartesian coordinates, was applied on the antenna substrate, as demonstrated in Figure 4a. In the structural analysis to evaluate the deformation of the antenna under this tensile strain, a linear static finite element analysis (FEA) was employed in this study using the commercial software ANSYS®. As the geometry and the applied displacements are also symmetric, while the material properties are assumed to be isotropic, symmetry boundary conditions were utilized, as shown in Figure 4b. In this study, it was assumed that the maximum applied tensile displacement on the stretchable antenna is 12 mm. The deformation of the stretchable antennas might be larger, but for demonstration purposes, 12 mm was considered in this study as the bound of the displacement of the stretchable antenna. The application can easily be extended for larger displacements in a future study.

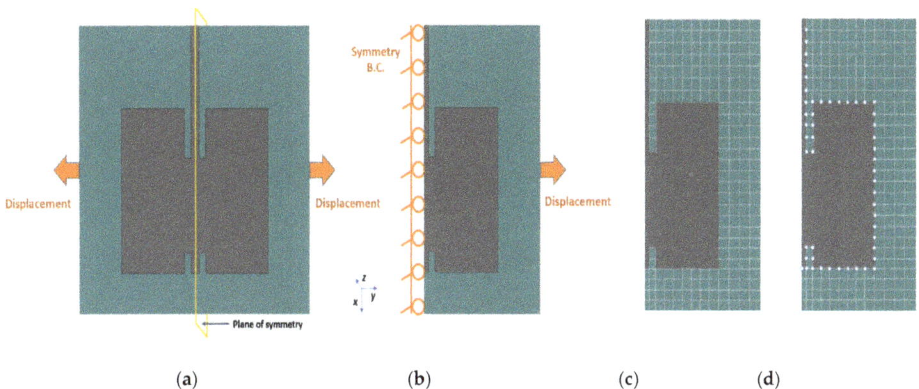

Figure 4. Schematic of the stretchable MSP antenna. (**a**) Boundary conditions for the tensile test; (**b**) symmetry conditions; (**c**) 202 divided substrate parts containing different thickness and 1 patch with a constant thickness; (**d**) 57 Cartesian coordinates of the patch.

The substrate of the antenna was discretized into 202 rectangular regions that represent the fluctuating substrate thickness, as shown in Figure 4c by white bounds. The outer geometry of the patch shown by gray color was discretized into 57 points, as depicted by white nodes in Figure 4c. Once the deformed shape of the antenna was obtained from the FEA of the antenna model in Figure 4, the return loss and R_f of the deformed antenna geometry were calculated using a commercial software called High Frequency Structure Simulator (HFSS).

4.1. Generation of Multivariate Data for the Varying Thickness

To model the substrate thickness fluctuation, a stochastic representation of a random field was taken into account. For this purpose, a Gaussian copula was employed to generate the random behavior corresponding to the thickness variation. Figure 5 demonstrates how the copula builds a joint distribution including two distinct sampling sets of thickness variation. First, the roughness of the substrate made of PDMS material was measured by Contour GT-I 3D Optical Microscope. The measured roughness data were added onto the mean thickness of the substrate, which is 0.9705 mm, to describe the variation of the thickness of the whole substrate. In this process, the Gaussian copula generated the random thickness samples engaging several marginal distributions of each sampling set of the thickness variation. Thus, the antenna can be modeled with the substrate, containing 202 small parts that have random different thicknesses and the patch with a thickness of 0.03 mm. Specifically, 500 FEAs were performed with different displacements generated within the range from 0 mm to 12 mm. For each of these simulations, 202 inter-correlated random thickness values were created from the Gaussian copula.

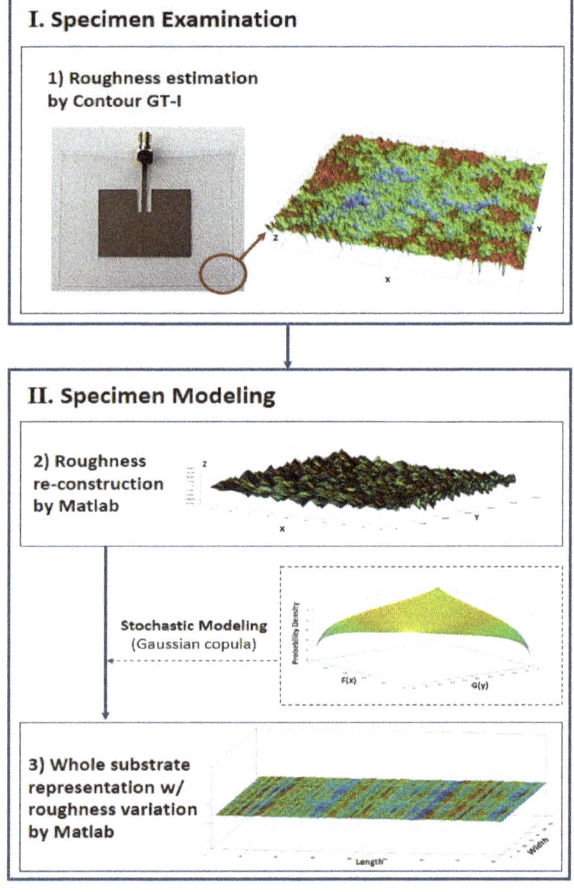

Figure 5. The Gaussian copula employed to demonstrate a joint distribution of random sets of thickness variation.

4.2. DR Process for Variation of the Substrate Thickness

Before conducting the DR, e of the initial thickness variation was calculated as 0.684 using Equation (16). This value is greater than 0.5, indicating that the correlation between the thickness data set is strong. This result indicates that FE should be used as the DR according to the proposed framework shown in Figure 3. The performances of two different FE methods, namely PCA and AE, were evaluated in this study. PCA and AE took 32 and 103 dominant components, respectively. In Table 2, the error was calculated for each method by

$$Error = mean\left(\sum\left(\frac{\left|D_{new} - D_{org}\right|}{\left|D_{org}\right|} \times 100\ (\%)\right)\right), \tag{17}$$

where D_{new} and D_{org} stand for the redundancy of diminished data and that of initial data, respectively. Unlike PCA, AE had higher redundancy because it took extra neuron training processes. However, the dimension taken by AE drew a lower reconstruction error (2.71%) than that taken by PCA (3.27%).

Table 2. Redundancy prediction and estimation error of the thickness coordinates. DR, dimension reduction; PCA, principal component analysis; AE, auto-encoder; ANN, artificial neural network.

		Redundancy Prediction		Coordinates Prediction Error		
	No DR	By PCA (Redundancy Reduction (%))	By AE (Redundancy Reduction (%))	By ANN (X coord./Y coord.)	By PCA/ANN (X coord./Y coord.)	By AE/ANN (X coord./Y coord.)
Thickness Coord.	11.053	5.503 (50.21%)	6.127 (44.57%)	7.03%/7.62 %	3.70%/5.41%	4.88%/7.09%

4.3. ANN-Based Surrogate Model to Predict Antenna Deformation

Once the DR was conducted, ANN was constructed to predict Cartesian coordinates of geometry points of the antenna patch, shown in Figure 4d, referring to the shape of the antenna under structural distortion. In the process, the number of variables of thickness variation and simulation was 202 and 500, respectively. The data obtained from these simulations were used to train the ANN model without the DR, with PCA, and with AE. As listed in Table 2, the trained ANN without DR led to a prediction inaccuracy of 7.03% and 7.62% for the x and y coordinates, respectively. When PCA and AE were used as two different DR methods, on the other hand, this error was reduced, as seen in the results in Table 2.

4.4. Dimension Reduction of Coordinates of the Patch

In Section 4.2, the DR was employed to analyze the antenna's mechanical behavior to obtain the coordinates of the geometry points of the deformed patch as the outputs. Then, these geometry points are used as the inputs for the electrical analysis of the deformed antenna to calculate the R_f value as the output. On the basis of the inputs, HFSS software was used to check the R_f behavior of the distorted antenna. As output data, 121 resonance frequencies were taken into account.

As this is a new analysis after the mechanical analysis, it was inevitable to conduct the DR for the electrical system behavior as well. The ECC value, e, was calculated as 0.442 for the deformed antenna inputs, which are the coordinates of the patch geometry points. This value was less than 0.5, indicating that FS should be used as the DR method in the electrical analysis step. The FS method used in this study was the IFT method and unnecessary coordinates were eliminated based on the significance value of 2. As a result of the IFT method, 28 meaningful coordinates were selected, as shown in Figure 6. Although the R_e value for the initial coordinates was calculated as 8.382, it was calculated as 5.623 for the deformed coordinates. Thus, it indicates that IFT precisely achieved the R_e reduction of 32.11% and the correlation of the original coordinates was changed to the independent and uncorrelated ones. In Figure 6, the white points represent the 28 points identified by the IFT method. It is seen that the number of input geometry points for the electrical analysis was reduced from 57 to 28. Although the total number of these selected points is still high, those are the necessary ones to accurately represent the geometry.

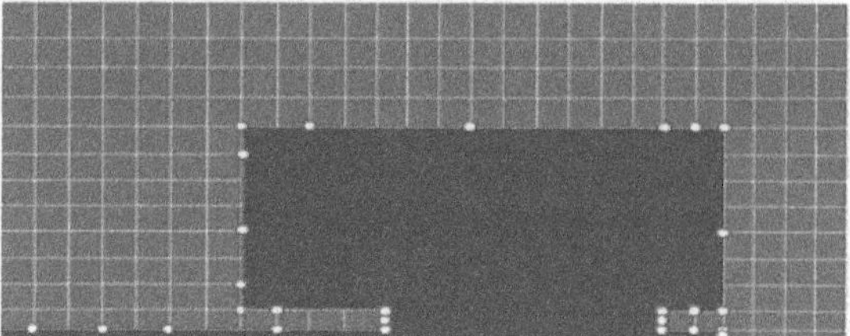

Figure 6. Selected new significant coordinates.

4.5. Optimization

Once the analyses were conducted for the initial geometry and the necessary geometry points of the antenna patch were determined, a shape optimization was utilized to improve the performance of the antenna. As explained in the introduction, diverse studies on the strain patch antenna have already been conducted by optimization, but most of them have concentrated on the sizing optimization to take updated height, width, or length of the antenna systems. The optimization may fail to boost the productivity of the antenna because the optimized design almost maintains its original shape. Topology optimization has also been employed to develop a new design, but its efficiency is only considerable under expensive additive manufacturing fabrication. In this study, therefore, a shape optimization was used to find the optimal shape of the antenna patch that minimizes the variation of the R_f and the return loss. It implies that, not only can the possibility of frequency shift induced by structural deformation be decreased, but also the performance of the antenna corresponding to electrical return loss will be maximized. Here, the lower return loss is guaranteed, and the better antenna performance is expected. Unfortunately, however, the explained general equations (Equations (1)–(5)) corresponding to the R_f variation have an assumption that the substrate should have a constant thickness. This assumption only draws a linear relation between the shift of the R_f and the strain. With respect to the stretchable strain MSP antenna including the thickness variation of the substrate, the direction of the R_f shifts could be anticipated by the general equations, but a correct calculation of the shift will not be guaranteed. For those reasons, in this study, a surrogate model to establish a relation between mechanical and electrical behavior of the antenna system was developed, engaging detailed FEA for structural deformation in ANSYS® and electrical analysis in HFSS. The R_f shift and the return loss respecting the structural deformation with additional attention of thickness variation were analyzed with the help of this surrogate model.

To establish objective functions, the substrate thickness variation and distorted patch's shape were considered. The function approximation modeled by ANN was combined with the shape optimization process. To escalate the capability of the antenna, an additional objective function was employed to truncate the return loss.

$$\text{Minimize} \quad \left[\frac{\left| R_{ff}(x, f_i) - R_{fs}(x, f_i, \varepsilon) \right|}{S(x, f_i, \varepsilon)} \right] \quad i = 1, 2. \tag{18}$$

$$\text{Subject to} \quad \left| R_{ff}(x, f_i) - R_{fs}(x, f_i, \varepsilon) \right| \leq \theta, \tag{19}$$

$$S(x, f_i, \varepsilon) + \gamma \leq 0, \tag{20}$$

$$x_l \leq x \leq x_u, \tag{21}$$

$$0 \leq \varepsilon \leq 30, \tag{22}$$

$$f_1 = 2.5\ \text{GHz} \ and \ f_2 = 5\ \text{GHz}, \tag{23}$$

where R_{ff} and R_{fs} stand for R_f of frequency samples and R_f under applied strain (ε). x, θ, S, and γ refer to a vector of the DVs, bandwidth regarding a return loss of -10 dB, the return loss, and acceptable return loss, respectively.

The bandwidth corresponding to -10 dB was employed to assure enough antenna operation, and the strain of 30% was established based on the boundary exploited in the FEA. This constraint enables linear static analysis. In order to utilize the benefit of UWB, dual bandwidths, 2.5 GHz and 5 GHz, which are preferred in a wireless fidelity (Wi-Fi) were considered. Moreover, a spline curve was used to model the geometry of the patch antenna with 28 geometry points. The spline coordinates were considered as the DVs, x, and the minimum and maximum boundaries of the coordinates were determined to bring optimum shape change of the initial MSP antenna. Particularly, the coordinates of feedline in the length direction were firmly restricted to keep the initial shape of the MSP antenna. The optimization process includes significant attention associated with the strain applied along the width direction. On the basis of the equation of the frequency shift, even though the strain occurs along the length direction, which can make the patch's length varied owing to the Poisson's ratio, the variation will not bring a major impact on the variation of the R_f. Thus, design space around the patch had much more design freedom to improve antenna's operation than that around the feedline. To conduct multidisciplinary design optimization, genetic algorithm (GA) was exploited in accordance with the developed ANN-based surrogate model. For the algorithm, 1000 iterations, crossover probability of 0.96, and mutation probability of 0.01 were used. After conducting the optimization process, a butterfly-shaped antenna was garnered, as shown in Figure 7. Locations of the acquired spline coordinates by the optimization were slightly modified to have a smooth contour. As anticipated, it focused on reducing the variations of the width of the antenna patch, not on that of the length. It could boost the antenna's operation under the applied longitudinal strain. The optimization process drew a reasonable outcome, coinciding with the equations mentioned above in Section 2.

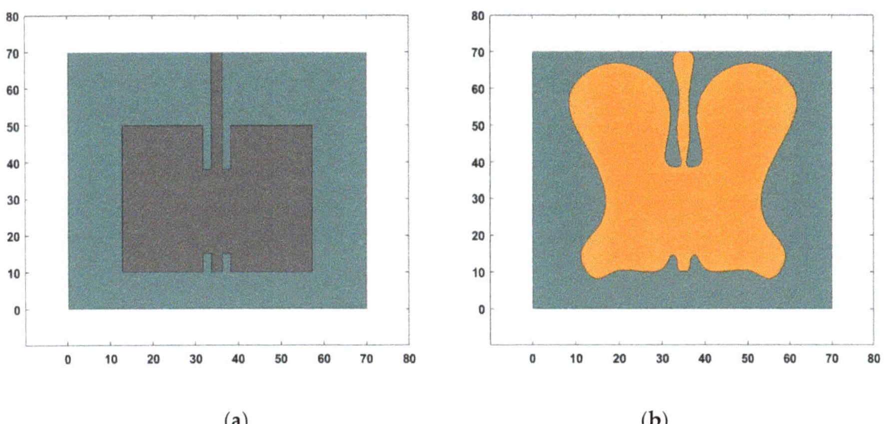

(a)   (b)

Figure 7. Design of the MSP patch antenna. (**a**) Initial antenna design; (**b**) optimal antenna design.

To assure the exactness of the optimized butterfly-shaped design, HFSS was utilized to evaluate the frequency shift. The results of the undeformed antennas are shown in Figure 8. In accordance with the employed bandwidth of -10 dB, an acceptable R_f range of the antenna was assigned from 2.3454 GHz (a) to 2.6539 GHz (b) and from 4.9130 GHz (c) to 5.1070 GHz (d) under 0 mm displacement, as depicted in Figure 8.

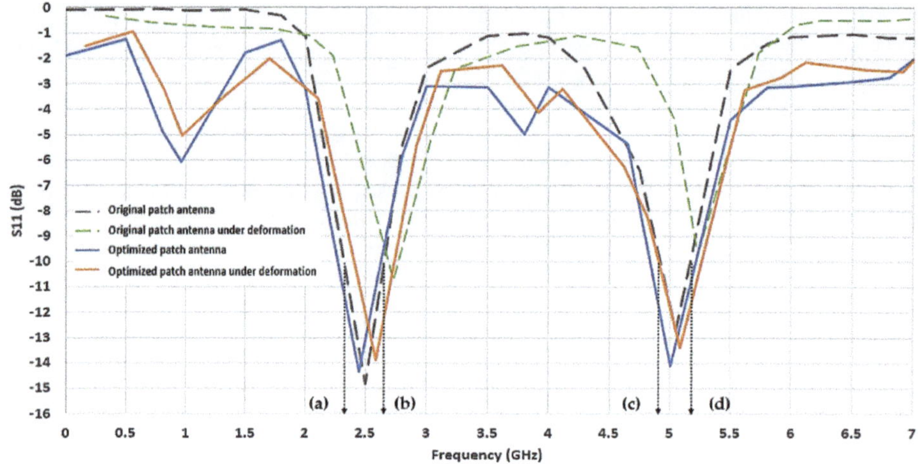

Figure 8. R_f comparison of initial, deformed, non-deformed optimal antenna, and deformed optimal antenna. (**a**) 2.3454 GHz; (**b**) 2.6539 GHz; (**c**) 4.9130 GHz; (**d**) 5.1070 GHz.

First, the capability of the suggested optimum design was compared with the initial antenna design when the deformation exists for both. Once the deformation of 12 mm is applied to the initial antenna, R_f drastically shifts, ranging from 2.5 GHz to 2.73 GHz and from 5.05 GHz to 5.23 GHz, which states that the antenna functionality will not be ensured because they deviate from each acceptable R_f range regarding 2.5 and 5.0 GHz. The return loss also decreased by 27.6% and 31.8%, respectively. As expected, the performance of the antenna was deteriorated under the deformation. Especially the antenna design in 5 GHz should be restricted owing to miserable return loss, which is higher than −10 dB. Contrary to the initial antenna, on the other hand, the optimized antenna excluding the DR process acquired credits for its flexible possibility under the tensile strain by enabling the R_f to be maintained in the reliable range (θ) and the return loss to be kept under −10 dB, as elucidated in Table 3.

Table 3. Comparison of resonance frequency (GHz) and return loss (dB) under strain test.

Estimated Design	Displacement of 0 mm				Displacement of 12 mm			
	Resonance Freq.		Return Loss		Resonance Freq.		Return Loss	
	at 2.5 GHz	at 5.0 GHz	at 2.5 GHz	at 5.0 GHz	at 2.5 GHz	at 5.0 GHz	at 2.5 GHz	at 5.0 GHz
Initial MSP antenna	2.5	5.05	−14.83	−12.73	2.73	5.23	−10.74	−8.68
Initial MSP antenna (with optimization)	2.45	5.02	−14.97	−13.64	2.59	5.09	−11.36	−11.20
Initial MSP antenna (with DR)	2.81	5.20	−14.11	−13.83	2.69	5.22	−11.07	−8.90
Initial MSP antenna (with DR and optimization)	2.48	4.74	−14.34	−14.11	2.65	5.10	−13.86	−13.37

Also, a comparison of two additional designs regarding the DR process antenna was conducted. In Table 3, the design of the initial MSP antenna design including DR led to a better operation than that of the initial antenna itself. Although, its improvement is not enough to dominate the initial antenna design with an optimization. As predicted, the optimized antenna design with DR had notable improvement when compared with all other cases. Therefore, the uncertainty-reduced optimal MSP antenna design could be a dominant approach because general wireless receivers would be employed without any additional extension work on the bandwidth.

4.6. PNN for Classification of the Resonance Frequency

In the final step, the reliability of the optimal MSP antenna was analyzed in accordance with the reliable range of θ. The antenna must operate within the range even though it is under deformation. The range was estimated by the non-deformed MSP antenna. From the previous step, the dual bands antenna under no displacement took the following range (θ):

$$2.3454 \text{ GHz} \leq \theta_{f\_2.5\text{GHz}} \leq 2.6539 \text{ GHz}$$
$$4.9130 \text{ GHz} \leq \theta_{f\_5.0\text{GHz}} \leq 5.1070 \text{ GHz} \quad ' \tag{24}$$

For reliable classification, the limit state function was exploited. According to the concept of the limit state function [53], the capacity or resistance of the antenna must be appointed to 2.5 GHz. However, if the absolute value of g is greater than 0.3085 (difference between 2.3454 GHz and 2.6539 GHz) and 0.194 (difference between 4.9130 GHz and 5.1070 GHz) at 2.5 GHz and 5 GHz, respectively, the designed strain MSP antenna will stay in class B, implying that the system must be ignored owing to the unreliable R_f. The Monte Carlo simulation (MCS) [53,54] was employed to estimate the probability of failure (P_f). On the basis of the P_f, the results of MCS with 10,000 random variables and PNN with 121 random variables were compared to assure results of the reliability estimation. As explained in Table 4, each P_f difference between PNN and MCS was 6.46% and 6.91% at 2.5 GHz and 5 GHz, respectively. In comparison with the original variables, the truncated coordinates obtained by IFT had a P_f of 0.291 and 0.325 by drawing 4.67% and 4.06% of P_f increase, respectively. Furthermore, the P_f acquired by IFT was close to that obtained by MCS (9.48% and 9.41%, respectively), which is less than 10%, so it can be a reliable threshold, eliminating computation-intensive tasks, for the classification process.

Table 4. Probability of failure conducted by probabilistic neural network (PNN) and Monte Carlo simulation (MCS).

	P_f of Original Data (2.5 GHz/5 GHz)	P_f of New Data by IFT (2.5 GHz/5 GHz)	P_f Difference of Original/New Data (2.5 GHz/5 GHz)
PNN	0.278/0.313	0.291/0.325	4.67%/4.06%
MCS	0.298/0.334	0.327/0.342	9.48%/9.41%
P_f difference of PNN/MCS	6.46%/6.91%	8.05%/5.14%	

5. Conclusions

In this research, a design framework exploiting a DR approach, machine learning-based surrogate modeling, structural optimization, and reliability assessment was proposed to handle a multivariate and multidisciplinary engineering system. The efficiency of the proposed design framework was addressed by a stretchable MSP antenna-based strain sensor.

Compared with the conventional rigid antenna, the efficiency of the proposed stretchable antenna design was highlighted because it contains careful consideration for all possible mechanical flexibility, which weights realistic system prediction and estimation. Such consideration demands meticulous examination on the flexible substrate that includes non-uniform thickness. In this research, therefore, the non-uniformity was regarded as the DVs to represent geometric uncertainty captured by a Gaussian copula function. With the copula function, a non-uniform substrate thickness was represented by 202 subparts, taking the mean of 0.9705 mm and the COV of 0.05. However, a huge number of DVs escalate the redundancy and complexity of data, inhibiting precise system prediction and estimation. In order to resolve the issue, the proposed framework employed a DR process, particularly FE and FS. In the DR process, ECC (e) was exploited as a clear guideline that suggests a better DR method grounded on data behavior. On the basis of the ECC estimation, FE was employed to reduce data redundancy regarding the substrate non-uniformity in which e is greater than 0.5, whereas FS eliminated unnecessary coordinates of the patch within the specific case ($0.5 \leq e < 1$). Both FE (PCA and AE) and FS (IFT) derived good redundancy reduction of 50.21%, 44.57%, and 32.11%, respectively. With the criterion, engineers can make a decision on the applicable DR process and management of multivariate data.

Moreover, owing to the limitation to formulate a relationship between structural deformation and electrical antenna's response, the ANN-based surrogate model employing the multivariate data purified by the DR process was developed to predict a complex engineering system. As the second drawback of the stretchable MSP strain sensor, the narrow bandwidth that restricts the functionality of the sensor was handled by a structural shape optimization. Compared with the conventional antenna shape, the proposed optimum shape drew the antenna's performance improvement of 5.77% at 2.5 GHz and 29.03% at 5 GHz within the reliable frequency range. A new optimum design implementing the DR process also escalated the performance improvement of 29.05% at 2.5 GHz and 35.08% at 5 GHz. Thus, it is confirmed that the developed optimal design maximized the frequency stability within a reliable bandwidth and antenna performance under structural deformation. In the final step, the reliability of the stretchable antenna for the strain sensor was assessed by PNN and validated by MCS. PNN evaluated the efficiency of the proposed design by showing the classification accuracy of 96%. The accuracy of PNN was validated by that of MCS. The results obtained for the stretchable strain MSP antenna show that the proposed design framework with the uncertainty characterization and dimension reduction is effective on multi-physics-based and multi-objective design processes.

Author Contributions: Conceptualization, S.-K.C.; methodology, S.H.; software, S.H.; validation, S.-K.C. and R.M.G.; formal analysis, S.H.; investigation, S.H.; resources, S.-K.C.; data curation, S.H.; writing—original draft preparation, S.H.; writing—review and editing, S.-K.C. and R.M.G.; visualization, S.H.; supervision, S.-K.C.; project administration, S.-K.C.; funding acquisition, S.-K.C. and H.-J.C. All authors have read and agreed to the published version of the manuscript.

Funding: This research was partially funded by Chung-Ang University Research Grants in 2016.

Conflicts of Interest: The authors declare no conflict of interest.

References

1. Ye, X.; Su, Y.; Han, J. Structural health monitoring of civil infrastructure using optical fiber sensing technology: A comprehensive review. *Sci. World J.* **2014**, *2014*. [CrossRef] [PubMed]

2. Liu, W.; Tang, B.; Han, J.; Lu, X.; Hu, N.; He, Z. The structure healthy condition monitoring and fault diagnosis methods in wind turbines: A review. *Renew. Sustain. Energy Rev.* **2015**, *44*, 466–472. [CrossRef]

3. Memmolo, V.; Elahi, H.; Eugeni, M.; Monaco, E.; Ricci, F.; Pasquali, M.; Gaudenzi, P. Experimental and Numerical Investigation of PZT Response in Composite Structures with Variable Degradation Levels. *J. Mater. Eng. Perform.* **2019**, *28*, 3239–3546. [CrossRef]

4. Zhu, M.; Xie, M.; Xu, W.; Cheng, L.K. A Nanocomposite-Based Stretchable Deformation Sensor Matrix for a Soft-Bodied Swallowing Robot. *IEEE Sens. J.* **2016**, *16*, 3848–3855. [CrossRef]

5. Li, T.; Tan, Y.; Shi, C.; Guo, Y.; Najdovski, Z.; Ren, H.; Zhou, Z. A High-Sensitivity Fiber Bragg Grating Displacement Sensor Based on Transverse Property of a Tensioned Optical Fiber Configuration and Its Dynamic Performance Improvement. *IEEE Sens. J.* **2017**, *17*, 5840–5848. [CrossRef]

6. Zhang, J.; Tian, G.Y.; Marindra, A.M.; Sunny, A.I.; Zhao, A.B. A review of passive RFID tag antenna-based sensors and systems for structural health monitoring applications. *Sensors* **2017**, *17*, 265. [CrossRef]

7. Hwang, S.; Choi, S.-K. Optimal Design of Stretchable Electronics With the Consideration of Response Variability. In Proceedings of the ASME 2017 International Design Engineering Technical Conferences and Computers and Information in Engineering Conference, Cleveland, OH, USA, 6–9 August 2017; p. V009T07A054.

8. Hwang, S.; Gorguluarslan, R.M.; Choi, S.-K.; Min, J.; Moon, J. Reliability Estimation of Stretchable Electronics Using a Dimension Reduction Framework. In Proceedings of the ASME 2016 International Design Engineering Technical Conferences and Computers and Information in Engineering Conference, Charlotte, NC, USA, 21–24 August 2016; p. V01AT02A025.

9. Lee, K.F.; Luk, K.M.; Lai, H.W. *Microstrip Patch Antennas*; World Scientific: River Edge, NJ, USA, 2017.

10. Chahat, N.; Zhadobov, M.; Sauleau, R. Antennas for body centric wireless communications at millimeter wave frequencies. In *Progress Compact Antennas*; IntechOpen: London, UK, 2014.

11. Jang, K.-I.; Han, S.Y.; Xu, S.; Mathewson, K.E.; Zhang, Y.; Jeong, J.-W.; Kim, G.-T.; Webb, R.C.; Lee, J.W.; Dawidczyk, T.J.; et al. Rugged and breathable forms of stretchable electronics with adherent composite substrates for transcutaneous monitoring. *Nat. Commun.* **2014**, *5*, 4779. [CrossRef]

12. Bosiljevac, M.; Casaletti, M.; Caminita, F.; Sipus, Z.; Maci, S. Non-uniform metasurface Luneburg lens antenna design. *IEEE Trans. Antennas Propag.* **2012**, *60*, 4065–4073. [CrossRef]

13. Chen, C.C.; Volakis, J.L. Bandwidth broadening of patch antennas using nonuniform substrates. *Microw. Opt. Technol. Lett.* **2005**, *47*, 421–423. [CrossRef]

14. Nakamoto, H.; Ootaka, H.; Tada, M.; Hirata, I.; Kobayashi, F.; Kojima, F. Stretchable strain sensor based on areal change of carbon nanotube electrode. *IEEE Sens. J.* **2015**, *15*, 2212–2218. [CrossRef]

15. Suikkola, J.; Björninen, T.; Mosallaei, M.; Kankkunen, T.; Iso-Ketola, P.; Ukkonen, L.; Vanhala, J.; Mäntysalo, M. Screen-printing fabrication and characterization of stretchable electronics. *Sci. Rep.* **2016**, *6*, 25784. [CrossRef] [PubMed]

16. Song, L.; Myers, A.C.; Adams, J.J.; Zhu, Y. Stretchable and reversibly deformable radio frequency antennas based on silver nanowires. *ACS Appl. Mater. Interfaces* **2014**, *6*, 4248–4253. [CrossRef] [PubMed]

17. Huang, Y.; Wang, Y.; Xiao, L.; Liu, H.; Dong, W.; Yin, Z. Microfluidic serpentine antennas with designed mechanical tunability. *Lab Chip* **2014**, *14*, 4205–4212. [CrossRef] [PubMed]

18. Kanaparthi, S.; Sekhar, V.R.; Badhulika, S. Flexible, eco-friendly and highly sensitive paper antenna based electromechanical sensor for wireless human motion detection and structural health monitoring. *Extrem. Mech. Lett.* **2016**, *9*, 324–330. [CrossRef]

19. Ramli, M.R.; Ibrahim, S.; Ahmad, Z.; Abidin, I.S.Z.; Ain, M.F. Stretchable Conductive Ink Based on Polysiloxane–Silver Composite and Its Application as a Frequency Reconfigurable Patch Antenna for Wearable Electronics. *ACS Appl. Mater. Interfaces* **2019**, *11*, 28033–28042. [CrossRef]

20. Xie, Z.; Avila, R.; Huang, Y.; Rogers, J.A. Flexible and Stretchable Antennas for Biointegrated Electronics. *Adv. Mater.* **2019**, *32*, 1902767. [CrossRef]

21. Vijayakumar, S.; D'souza, A.; Schaal, S. Incremental online learning in high dimensions. *Neural Comput.* **2005**, *17*, 2602–2634. [CrossRef]

22. Chandrashekar, G.; Sahin, F. A survey on feature selection methods. *Comput. Electr. Eng.* **2014**, *40*, 16–28. [CrossRef]

23. Kang, M.; Islam, M.R.; Kim, J.; Kim, J.-M.; Pecht, M. A hybrid feature selection scheme for reducing diagnostic performance deterioration caused by outliers in data-driven diagnostics. *IEEE Trans. Ind. Electr.* **2016**, *63*, 3299–3310. [CrossRef]

24. Lei, Y.; Jia, F.; Lin, J.; Xing, S.; Ding, S.X. An intelligent fault diagnosis method using unsupervised feature learning towards mechanical big data. *IEEE Trans. Ind. Electr.* **2016**, *63*, 3137–3147. [CrossRef]

25. Durante, F.; Sempi, C. *Principles of Copula Theory*; Chapman and Hall/CRC: London, UK, 2015.

26. Shlens, J. A tutorial on principal component analysis. *arXiv* **2014**, arXiv:1404.1100.

27. Witten, I.H.; Frank, E.; Hall, M.A.; Pal, C.J. *Data Mining: Practical Machine Learning Tools and Techniques*; Morgan Kaufmann: Burlington, MA, USA, 2016.

28. Roiger, R.J. *Data Mining: A Tutorial-Based Primer*; Chapman and Hall/CRC: London, UK, 2017.

29. Taylor, J.D. *Ultra-Wideband Radar Technology*; CRC Press: Boca Raton, FL, USA, 2018.

30. Singh, V.; Ali, Z.; Ayub, S.; Singh, A. Bandwidth optimization of compact microstrip antenna for PCS/DCS/bluetooth application. *Open Eng.* **2014**, *4*, 281–286. [CrossRef]

31. Sato, Y.; Campelo, F.; Igarashi, H. Fast shape optimization of antennas using model order reduction. *IEEE Trans. Magn.* **2015**, *51*, 1–4. [CrossRef]

32. Wang, J.; Yang, X.-S.; Ding, X.; Wang, B.-Z. Topology Optimization of Conical-Beam Antennas Exploiting Rotational Symmetry. *IEEE Trans. Antennas Propag.* **2018**, *66*, 2254–2261. [CrossRef]

33. Bhatt, S.; Mankodi, P.; Desai, A.; Patel, R. Analysis of ultra wideband fractal antenna designs and their applications for wireless communication: A survey. In Proceedings of the 2017 International Conference on Inventive Systems and Control (ICISC), Coimbatore, India, 19–20 January 2017; pp. 1–6.

34. Xu, R.; Jang, K.I.; Ma, Y.; Jung, H.N.; Yang, Y.; Cho, M.; Zhang, Y.; Huang, Y.; Rogera, J.A. Fabric-based stretchable electronics with mechanically optimized designs and prestrained composite substrates. *Extreme Mech. Lett.* **2014**, *1*, 120–126. [CrossRef]

35. Shapira, J. Active Antenna Array Configuration and Control for Cellular Communication Systems. U.S. Patent 20030073463A1, 31 May 2005.

36. Chen, S.-C.; Wang, Y.-S.; Chung, S.-J. A decoupling technique for increasing the port isolation between two strongly coupled antennas. *IEEE Trans. Antennas Propag.* **2008**, *56*, 3650–3658. [CrossRef]

37. Har-Peled, S.; Indyk, P.; Motwani, R. Approximate nearest neighbor: Towards removing the curse of dimensionality. *Theory Comput.* **2012**, *8*, 321–350. [CrossRef]

38. Sricharan, K.; Wei, D.; Hero, A.O. Ensemble estimators for multivariate entropy estimation. *IEEE Trans. Inf. Theory* **2013**, *59*, 4374–4388. [CrossRef]

39. Ince, R.A.; Giordano, B.L.; Kayser, C.; Rousselet, G.A.; Gross, J.; Schyns, P.G. A statistical framework for neuroimaging data analysis based on mutual information estimated via a gaussian copula. *Hum. Brain Mapp.* **2017**, *38*, 1541–1573. [CrossRef]

40. Guyon, I.; Gunn, S.; Nikravesh, M.; Zadeh, L.A. *Feature Extraction: Foundations and Applications*; Springer: Berlin, Germany, 2008.

41. Tang, J.; Alelyani, S.; Liu, H. Feature selection for classification: A review. In *Data Classification: Algorithms and Applications*; Springer: Berlin, Germany, 2014; p. 37.

42. Hinton, G.E.; Salakhutdinov, R.R. Reducing the dimensionality of data with neural networks. *Science* **2006**, *313*, 504–507. [CrossRef]

43. Melingui, A.; Lakhal, O.; Daachi, B.; Mbede, J.B.; Merzouki, R. Adaptive neural network control of a compact bionic handling arm. *Mechatron. IEEE ASME Trans.* **2015**, *20*, 2862–2875. [CrossRef]

44. Höhndorfa, L.; Sembiringa, J.; Holzapfela, F. Copulas applied to Flight Data Analysis. In Proceedings of the Probabilistic Safety Assessment and Management PSAM, Honolulu, HI, USA, 22–27 June 2014; Volume 12.

45. Eid, H.F.; Hassanien, A.E.; Kim, T.; Banerjee, S. Linear correlation-based feature selection for network intrusion detection model. In *International Conference on Security of Information and Communication Networks*; Springer: Sydney, Australia, 2013; pp. 240–248.

46. Numata, J.; Ebenhöh, O. Measuring correlations in metabolomic networks with mutual information. *Genome Inf.* **2008**, *20*, 112–122.

47. Nga, N.T.T.; Khanh, N.K.; Hong, S.N. Entropy based correlation clustering for wireless sensor network in multi-correlated regional environment. In Proceedings of the 2016 International Conference on Electronics, Information, and Communications (ICEIC), Da Nang, Vietnam, 27–30 January 2016; pp. 1–4.

48. Min, J.H.; Lee, Y.-C. Bankruptcy prediction using support vector machine with optimal choice of kernel function parameters. *Expert Syst. Appl.* **2005**, *28*, 603–614. [CrossRef]

49. Nock, R.; Nielsen, F. Bregman divergences and surrogates for learning. *IEEE Trans. Pattern Anal. Mach. Intell.* **2008**, *31*, 2048–2059. [CrossRef] [PubMed]

50. Han, J.; Pei, J.; Kamber, M. *Data Mining: Concepts and Techniques*; Elsevier: Burlington, MA, USA, 2011.

51. Mason, C.; Twomey, J.; Wright, D.; Whitman, L. Predicting engineering student attrition risk using a probabilistic neural network and comparing results with a backpropagation neural network and logistic regression. *Res. High. Educ.* **2018**, *59*, 382–400. [CrossRef]

52. Arnaiz-González, Á.; Fernández-Valdivielso, A.; Bustillo, A.; de Lacalle, L.N.L. Using artificial neural networks for the prediction of dimensional error on inclined surfaces manufactured by ball-end milling. *Int.J. Adv. Manuf. Technol.* **2016**, *83*, 847–859. [CrossRef]

53. Choi, S.-K.; Grandhi, R.; Canfield, R.A. *Reliability-Based Structural Design*; Springer Science & Business Media: Berlin, Germany, 2006.

54. Coro, A.; Abasolo, M.; Aguirrebeitia, J.; de Lacalle, L.L. Inspection scheduling based on reliability updating of gas turbine welded structures. *Adv. Mech. Eng.* **2019**, *11*, 1687814018819285. [CrossRef]

 *applied sciences*

Article

Product Service System Availability Improvement through Field Repair Kit Optimization: A Case Study [†]

Eun Suk Suh

Graduate School of Engineering Practice, Institute of Engineering Research, Seoul National University, Seoul 08826, Korea; essuh@snu.ac.kr; Tel.: +82-2-880-7175

† This article is a re-written and extended version of "Improving Product Service System Availability through Field Repair Kit Optimization" presented at the 2nd East Asia Industrial Engineering Workshop (EAWIE 2015), Seoul, Korea, on 7 November 2015.

Received: 11 September 2019; Accepted: 9 October 2019; Published: 12 October 2019

Abstract: Product service system (PSS) is becoming a popular business model, where companies offer product based service to customers to realize steady recurring revenue. However, to provide PSS-based service to customers in reliable way, PSS need to be supplemented with a field repair kit onsite, in case of parts failure and PSS shutdown. The field repair kit consists of frequently used spare parts in multiple quantities. However, mismatch in spare parts type and quantities in the field repair kit will results in sub-par performance of PSS for both customer and company. In this paper, a case study involving industrial PSS repair kit optimization is presented. In the case study, the field repair kit for complex industrial printing system is cost optimized, while satisfying the system availability requirement, specified by the maintenance contract between the company and the customer. Key analysis steps and results are presented to offer insight into the PSS field repair kit optimization, offering useful references to industrial practitioners.

Keywords: product service system (PSS); availability; field repair kit

---

1. Introduction

Research on the product service system (PSS) [1,2] is increasing as many leading global industries are shifting their focus toward product based services. For example, in the aircraft engine industry, the old business model aimed to sell an engine to an aircraft manufacturer and provide engine maintenance, if necessary. Over time, this business model has evolved into a service centric model, often called the "power by the hour" approach [3], where aircraft engine companies own the engines installed on customer's aircrafts, and charge their customers for actual flight hours. This paradigm shift impacted many facets of product development and lifecycle management, including product requirement definition, subsystem design, service development, and total product lifecycle cost analysis.

According to general literature survey by Beuren [1], academia defines PSS as "a combination of products and services in a system that provides functionality for consumers and reduces environmental impact" [4]. Similarly, Baines et al. [5] defined PSS as "an integrated product and service offering that delivers value in use to the customer". Tukker [6] further categorized PSS into eight different types, which are clustered into product-related services, use-oriented services, and result-oriented services. The PSS presented in our case study is the product-related service type, whose company offers a product and related services (e.g., financing, maintenance contract, spare and consumable parts supply) throughout the lifecycle of the product.

There are four different essential elements in the PSS: product, process, related human role, and service. Product typically consists of actual hardware equipment used to provide specific functions required. Process is the sequence of actions required to provide the desired service. A related human role consists of activities that need to be performed by personnel who are part of the PSS. Finally, the service is the output of all these equipment and activities that is provided to customers. One such example of PSS is an amusement park ride. The product is the park ride itself, which is necessary for providing necessary entertainment service to ride users. The process is a sequence of activities necessary to provide ride service, which consists of getting customers into the ride, operating the ride equipment, getting customers out of the ride, and equipment maintenance. Related human role include operating and maintaining ride equipment. The service aspect is the ride experience itself, enjoyed by customers.

For PSS, one of the key performance metrics is the system availability. The PSS must be operational to meet the availability requirement promised to customers. For the amusement park ride example previously mentioned, the ride apparatus must be operational during park business hours to provide customers with a satisfactory ride experience. Not meeting this availability requirement, due to ride apparatus failure, will result in loss of customers and revenue. One way to maximize the uptime availability of the PSS is to keep a field repair kit, consisting of spare parts for the PSS, on the customer site, in case of such a failure. In many cases, product design teams rely on their prior experiences to determine the types and quantities of spare parts in field repair kits. However, this often result in a sub-optimal field repair kit, containing some spare parts that are never used, or carrying less-than-necessary quantities of some spare parts that are always used. This mismatch of spare parts inventory may cause unacceptable downtime and lost revenue for PSS customers and providing companies.

In this paper, a PSS field repair kit optimization case study is presented in detail. In the case study, a field repair kit for a complex PSS (industrial printing system) was optimized in terms of field inventory kit cost, while satisfying the availability requirement set by contract with the customer. A high fidelity simulation PSS simulation model was created to simulate spare parts usage during the PSS operation. Using the model, a cost optimal field repair kit was identified. Subsequent analysis was performed to determine the confidence level that the PSS is capable of achieving the imposed PSS availability requirement. The paper is organized as follows. A survey of related research literature is presented in Section 2. The case study overview, optimization process, analysis of the results, and discussion are presented in Section 3. The paper closes with conclusions in Section 4.

2. Literature Review

Traditionally, product manufacturers have focused their primary efforts on product development and sales. For these traditional manufacturing firms, the term "service" consisted of maintenance and repair of their products in the field as needed. However, as companies seek ways to create recurring revenues throughout the lifecycle of their products, the trend to integrate the product with the services offered to the customer has increased. This approach has advantage of continuous and stable revenue generation for the company because it establishes a long term relationship between the company and the customer [7], as well as sustaining environment from the social and company perspectives [8,9]. There have been successful examples of such practices, which include Gage Products and PPG Industries [10], where companies have integrated their core products with the total service offered to the customer. For example, major printing systems manufacturing companies, such as Xerox and HP, offers total document management services to various customers to optimize their total document production workflows.

Once the PSS is deployed in the field, it is of critical importance that the deployed PSS is operational, meeting the contracted availability requirement promised to the customer. One way to meet the availability requirement is through implementing appropriate maintenance policy, which includes preventive maintenance and predictive maintenance. There have been several works published in the field of preventive maintenance and predictive maintenance. Relevant works on preventive maintenance include work by Adhikary et al. [11], who proposed multi-objective genetic algorithm approach to optimize availability and maintenance cost through preventive maintenance scheduling model, work by Moghaddam [12], who proposed a nonlinear mixed-integer optimization model for a manufacturing system, and the work by Mokhtar et al. [13], who proposed a maintenance policy optimization framework using Bayesian networks. Additionally, related works on predictive maintenance include work by Van Horenbeek and Pintelon [14], who proposed a dynamic predictive maintenance policy and compared it with other various maintenance policies to show that the proposed policy results in significant cost reduction. Another work by Wang et al. [15] proposed cloud-based predictive maintenance paradigm to advance the state of intelligent manufacturing. Finally, there is a research trend that aims to link big data to improve predictive maintenance [16,17].

Another way to achieve required PSS availability is through implementing component redundancy and keeping an onsite spare parts field repair kit, consists of parts that are expected to fail [18]. This will prevent the unintended PSS downtime due to lack of critical spare parts. However, deciding on the quantity of each spare part of the PSS field repair kit requires information on the total PSS life cycle, average usage per designated period, and individual spare part's reliability. The management of spare parts inventory is a well-developed research topic. Published works have investigated topics such as the allocation of spare parts inventory within a multi-echelon supply chain [19–21], spare parts inventory and reliability decision framework with service constraints [22,23], combined optimization of preventive maintenance and spare parts inventory [24,25], and obsolescence management [26]. Recent advances in PSS-related research has expanded to other important areas, such as sustainable product service system design [27,28], cloud based product service system design [29,30], product service system implementation for the smart city [31], and incorporation of digital twin concept to the product service systems [32]. Recent works on field repair kit optimization include the joint planning and optimization of spare parts inventory and service engineer staffing [33,34].

In this paper, the featured case study focuses on the cost optimization of the onsite spare parts repair kit, subject to the PSS availability requirement specified by the contract between the company and the customer. Once the field repair kit is optimized for a specific level of availability requirement, additional analysis will determine the level of confidence of the PSS in meeting the availability requirement with the repair kit. The work presented in this study contributes to existing literature by introducing a real industry PSS case study that can serve as a reference for other industry practitioners.

3. Product Service System Case Study

3.1. Case Study Overview

Xerox, one of the world's leading printing system manufacturers, has a fleet of industrial printing systems deployed in the field. Industrial printing systems are installed at customer's print shops, where they become part of production systems to produce printing goods, which, in turn, are sold to their customers. Xerox (the company) and the customer typically have a recurring fee based maintenance contracts, in which the company maintains customer's printing system on regular basis, supplies necessary consumable parts required for operation, and performs repairs in case of printing system failure. Usually, as part of the maintenance contract, the company is responsible to have its printing system available for production above a certain agreed threshold during the regular operating hours. Failure by the company to meet the contracted system availability requirement may result in penalty to the company, or even worse, cancellation of the maintenance contract by the customer. In that sense, the industrial printing system can be classified as a PSS for Xerox, and will be referred to as such in subsequent sections.

Since unexpected system failures are inevitable, the company supplies customers with a field repair kit, consists of frequently replaced spare parts. Typically, the composition of the spare parts and the quantity of each spare part in the field repair kit are determined by printing system design team, based on historical performance of similar parts in other previously launched products. After the product launch, the composition and quantities of spare parts in the field repair kit are updated as the true life of spare parts become known.

However, on many occasions, this result in a sub-optimal field repair kit, where some parts are never used due to less than expected parts failure, while some parts are in constant shortage and demand as they are used more frequently than expected. This mismatch creates an undesirable situation for both the company and customers. Customers experience more unexpected PSS downtimes, above the threshold established by the maintenance contract, due to a shortage of critical spare parts. This will impact a customer's ability to meet their business goals. As for the company, mismatch in field repair kit spare parts type and quantity may result in their failure to meet the availability requirement promised to the customer, jeopardizing their credibility, and sometimes terminating their contract with that particular customer. Another issue concerns the field repair kit inventory itself. In some cases, the company financially owns the field repair kit inventory, tying up corporate capital in the form of spare parts inventory. If some spare parts become obsolete while in a company's possession, the company would incur unnecessary financial loss due to the extra cash that was tied up in the form of unused parts, and the additional cost for management of obsolete parts. The constant shortage of some spare parts can create an extra cost issue in terms of expedited shipment from parts suppliers and to customer's sites. In this case study, a simulation based optimization framework was created and utilized to configure a cost-optimal PSS field repair kit to satisfy the specific availability requirements. One of several industrial printing system produced by the company was used as a specific example.

To optimize the PSS field repair kit and to perform further analysis, following steps were followed. First, a high fidelity PSS simulation model was created to simulate the system operation and availability, subject to spare parts failure. Next, using the simulation model, a series of key parameter variation experiments were conducted to yield PSS availability under specific parameter settings. The obtained results were used to construct a regression based PSS availability model that would describe PSS availability as the function of key parameters. Regression model was then used to optimize the field repair kit in terms of the inventory cost, while satisfying the availability requirement. Finally, a Monte Carlo simulation of PSS with the optimized field repair kit was performed to estimate the confidence level for the PSS to meet the specific availability requirement.

3.2. Case Study Assumptions

Several assumptions were made to establish the boundary within which the case study is valid, and to reflect reality. Simulations, subsequent optimizations, and analysis were performed within the framework of the assumptions.

• Definition of availability: in this case study, the availability of the PSS is defined as follows. When the PSS shuts down due to a spare part failure, the part is replaced with a fresh part from the field repair kit. If there is no replacement part present in the field repair kit inventory, it must be shipped from the central warehouse. This is recorded as a PSS down incident. The availability metric is the percentage of all PSS part requests that is fulfilled by immediate replenishment from the field repair kit. For example, if there were 100 requests for various spare parts, and 95 of them were fulfilled from the field repair kit without placing emergency order from the central warehouse, the availability of the PSS is 95%.

• PSS and production system configuration: for the case study, the production system consists of two PSSs of the same type in a serial configuration. The field repair kit optimized in the case study is configured to service both PSSs.

• Field repair kit spare parts composition and quantities: the design team provided the list of spare parts in the field repair kit, along with their respective quantities and the estimated life. For the case study, all parts identified by the design team were included in the optimization. A total of 55 spare parts were included in the field repair kit.

• Periodic PSS usage: the PSS usage is defined as the number of prints produced per time period (month in this case study). For the case study, the nominal usage is set to 8.0 million, as provided by the design team responsible. The usage data was derived from past historical data of customers who have used similar type of PSS. Additionally, it was assumed that all spare parts fail based on PSS usage, not based on time.

• Onsite conditions: availability of PSS also depends on other aspects, such as experience of PSS operation and maintenance staff and production equilibrium. It was assumed that the level of expertise of operation and maintenance staff was competent to upkeep the PSS at the desired performance level, as it typically is for real onsite facilities. Additionally, it was assumed that the production equilibrium is maintained.

• Simulation assumption: for the Monte Carlo simulation, spare parts life distribution and periodic PSS usage distribution were provided by the design team. Additionally, simulation time was set to be equal to the typical contract term agreed between the company and the customer. For this case study, the contract term was set to five years, with a single simulation period equal to one month.

3.3. Product Service System Simulation Model

The first step in the proposed process is to create a high fidelity PSS model. The Anylogic™, software was used to create the PSS model to simulate PSS usage and usage-based spare parts failures. Figure 1 shows the basic structure of the PSS discrete event simulation model. The PSS selected for the case study is typically sold in pairs (quantities of two) to customers, and used in serial configuration to process customer orders for the duration of the contract period. There are work in process (WIP) product inventory as shown in the model. When the simulation starts, document orders start to arrive into the facility. Each order contains a request for certain quantities of a finished product, which, in this case, is the document produced in the desired quantity. The production line initiates product manufacturing, and utilizes the PSS that is part of the total production system. Once requested product quantity is manufactured, the order is complete and the finished document products are processed for delivery to final recipients. The simulation model records PSS usage, which is equal to the total quantity of the products manufactured. When the PSS usage reaches a threshold for a particular spare part with a specific usage based failure rate, the part fails, and the PSS shuts down. The failed part is replaced with the part from the field repair kit inventory. Once the spare part is replaced, the usage count for that specific spare part is reset and the production resumes. The replaced part is replenished from the central warehouse to fill the field repair kit. If the failed part is out of stock at the customer's site, then an emergency order is placed, and the part is shipped from the central warehouse immediately, using one-day express shipment. When this happens, it is recorded as a down incident for the PSS.

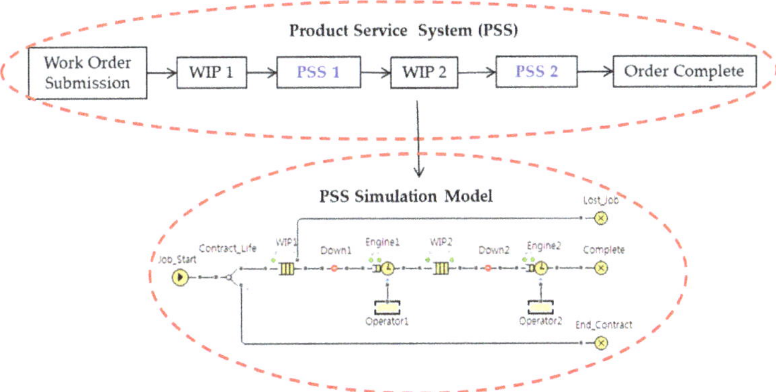

Figure 1. Simulation model flow chart for industrial printing product service system.

Figure 2 shows the Input-Process-Output (IPO) diagram for the PSS simulation model shown in Figure 1. Required input data are spare part life, quantity of spare part onsite and the monthly usage of PSS. Once the simulation ends, the model yields three specific outputs: (1) total demand for the spare part required by PSS, (2) the total number of PSS down incidents due to spare part shortage, and (3) the PSS availability.

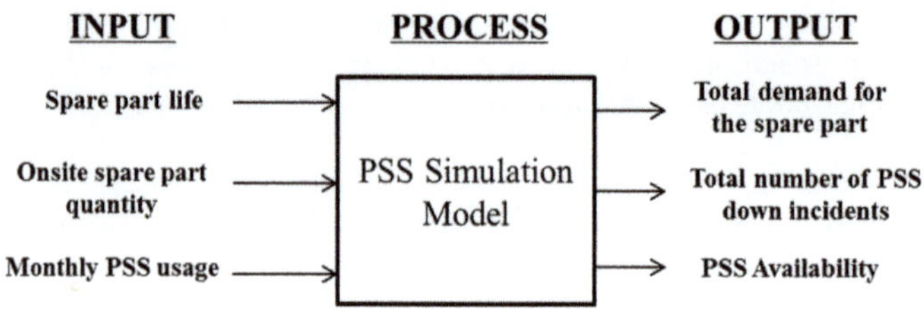

Figure 2. Input-Process-Output (IPO) diagram of the simulation model.

3.4. PSS Availability Model

The second step in the overall process is to construct the PSS availability regression model by performing key parameter variation experiments for PSS using the simulation model. A model of PSS availability as a function of input parameters shown in Figure 2 is created. Table 1 shows the key parameters and ranges of the parameter values used for the case study.

To create a reliable model, 240 different experiment runs were set up and performed using different deterministic parameter values, and resulting PSS availability values were recorded. Table 2 shows results of the 15 deterministic experiment runs out of 240 runs performed. For example, experiment number 2 in Table 2 shows a spare part with a life of 10 million uses (0.1 failure per million use), with only one provided in the field repair kit. When the simulation was performed with PSS monthly usage of 10.2 million, the availability of the PSS overall was 15% due to shortage of the part. Availability for each run was recorded. Once all experiment runs are complete, a regression based PSS availability model was constructed using parameters shown in Table 1. The regression model provides a good surrogate means to calculate PSS availability using spare part failure rate, onsite spare part quantity, and PSS usage per period. One thing to note is that the regression model is valid within parameter value ranges shown in Table 1.

Table 1. Key parameters and their value ranges for simulation and optimization.

Parameters	Values Ranges
Spare part failure rate (1/spare part life)	0.001–0.5 failures per million prints
Onsite spare part quantity	0–10
PSS usage per unit period	5.7–10.2 million prints

Table 2. Partial results of deterministic PSS simulation.

Runs	Spare Part Failure Rate (per Million Prints)	Quantity in the Field Repair Kit	PSS Availability based on Monthly PSS Usage		
			5.7 million	8.0 million	10.2 million
1	0.032	2	100%	100%	78%
2	0.100	1	50%	50%	15%
3	0.100	5	100%	100%	56%
4	0.500	2	100%	100%	80%
5	0.500	10	100%	100%	88%

3.5. Cost Optimization Model

For the next step of the process, the PSS availability regression model created from the previous step is embedded into the field repair kit cost optimization model. The IPO diagram for the optimization model is shown in Figure 3.

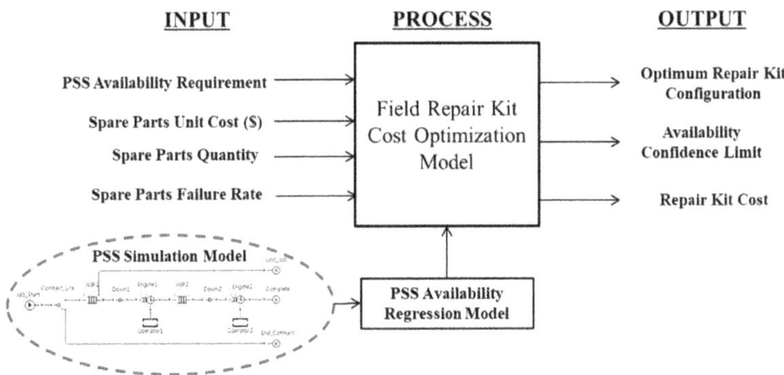

Figure 3. IPO diagram of the field repair kit cost optimization model.

For the field repair kit cost optimization, several input parameters are required. The first parameter is the required PSS availability target. This is typically dictated by the customer who utilizes the PSS when the maintenance contract is signed. The second parameter is the unit cost for each spare part in the field repair kit. The third parameter is the quantity of each spare part in the field repair kit, and the last parameter is the failure rate of each spare part. Table 3 shows a partial list of the actual spare parts included in the field repair kit, with actual parameter values. It should be noted that the unit cost of each spare part is added, since the objective of the optimization is to minimize the total cost of the field repair kit. For the monthly PSS usage, it was set at 8.0 million, based on field data for similar systems. Optimization was performed using spreadsheet-based commercial software (QuantumXL™), which uses a heuristic algorithm.

Table 3. Partial list of spare parts and quantities in the PSS field repair kit originally proposed by the design team.

Spare Part	Unit Cost (Normalized)	Original Part Quantity in the Field Repair Kit	Failure Rate (per Million Prints)
A	100.00	2	0.033
B	2.89	1	0.040
C	1.23	1	0.035
D	0.78	6	0.500
E	0.77	1	0.035
F	0.29	2	0.033
G	0.19	1	0.050
H	0.17	4	0.056
I	0.05	4	0.040
J	0.03	4	0.056
K	0.01	2	0.500

For the optimization, three different scenarios were formulated after discussion with the design team. Table 4 lists proposed scenarios. The first scenario serves as the base scenario, where the PSS operates with the original field repair kit composition shown in Table 3. In the second scenario, the PSS availability requirement parameter is set to 95%, and the field repair kit is optimized for total cost. For the third scenario, the PSS availability requirement parameter is relaxed to 90%.

Table 4. Key parameters and their value ranges for simulation and optimization.

Scenario	Scenario Description
1: Base	PSS field repair kit composition is provided by the PSS design team.
2: 95% availability	PSS field repair kit is cost optimized with PSS availability requirement set to 95%.
3: 90% availability	PSS field repair kit is cost optimized with PSS availability requirement set to 90%.

3.6. Optimization Results and Discussion

Table 5 shows a partial list of the optimized field repair kit spare parts shown in Table 3. The list for scenario 1 shows the original quantities of spare parts proposed by the design team. Second column shows the optimized spare parts list for 95% availability requirement. The last column shows a partial list of spare parts, optimized for 90% availability requirement. Observation of the list reveals some key insights. In the original repair kit composition, two units of spare part A were kept on the customer site. However, the optimum composition eliminated spare part A altogether from the field repair kit, mainly because of the part unit cost, which was significantly higher than the cost of any other part in the field repair kit. On the other hand, part J, which has a very low part unit cost, saw an increase in quantity from the original quantity proposed by the design team. The observation of the optimization results is that in balancing the field repair kit cost and the availability requirement, the optimized field repair kit is composed in a way that the PSS will only shut down for a part that is too expensive to be held onsite, and not for a low cost part.

Table 5. Partial list of spare parts and their quantities in the optimized field repair kit for each scenario.

Spare Part	Unit Cost (Normalized)	Spare Parts Quantity		
		Scenario 1	Scenario 2	Scenario 3
A	100.00	2	0	0
B	2.89	1	5	4
C	1.23	1	12	4
D	0.78	6	12	10
E	0.77	1	16	7
F	0.29	2	19	11
G	0.19	1	14	6
H	0.17	4	21	6
I	0.05	4	21	11
J	0.03	4	19	19
K	0.01	2	20	14

Figure 4 shows, for each scenario, the total of the cost of the PSS field repair kit. The results show that in order to meet the PSS availability requirement of 95%, the total cost of the field repair kit must increase by more than 60% over that of the originally proposed field repair kit composition in Scenario 1. However, when the PSS availability requirement constraint is relaxed to 90%, the cost of the field repair kit was significantly reduced, nearly down to 50% of the field repair kit cost shown in Scenario 2. In other words, increasing the availability requirement by 5% resulted in a 50% increase in the total cost of the field repair kit. Additionally, it is observed that if 90% PSS availability is acceptable to the customer, the optimized field repair kit can be composed for lesser cost than the original field repair kit proposed by the design team.

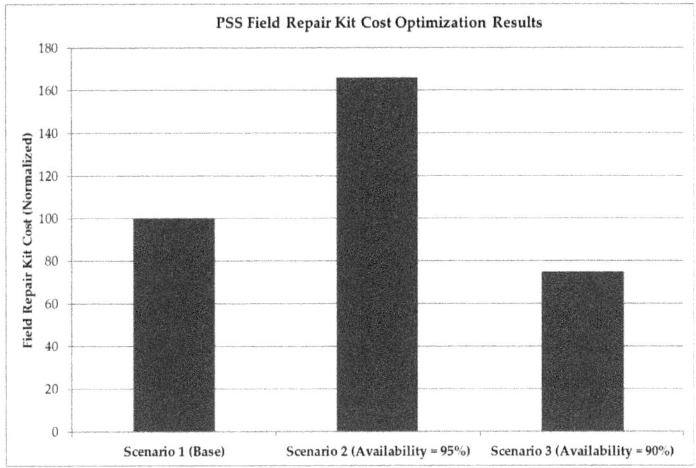

Figure 4. Field repair kit cost optimization results for scenarios analyzed.

As in the last step, using the optimized field repair kit composition, Monte Carlo simulation was performed to estimate the confidence level for which the PSS can meet the availability requirement for each scenario. Two parameters were used. The first parameter was the failure rate of each spare part. Distribution of spare parts failure rates were set based on the historical reliability data of the part or similar parts. The second parameter was the PSS usage. Since PSSs are used by many customers, but with different monthly usage, the PSS monthly usage was set with triangular distribution with a minimum value of 5.7 million uses and a maximum value of 10.2 million uses, with a mean usage of 8.0 million per month. This was based on the estimate from the design team. The resulting distribution of availability requirement was analyzed to determine the confidence level for the optimized field repair kit to meet the target PSS availability. The confidence level of availability that the PSS can achieve with the optimized field repair kit is plotted in Figure 5. For each scenario, 10,000 simulation runs were performed with the spare parts failure rate distribution and the periodic PSS usage distribution. The horizontal axis represents the confidence level, while the vertical axis represents the PSS availability.

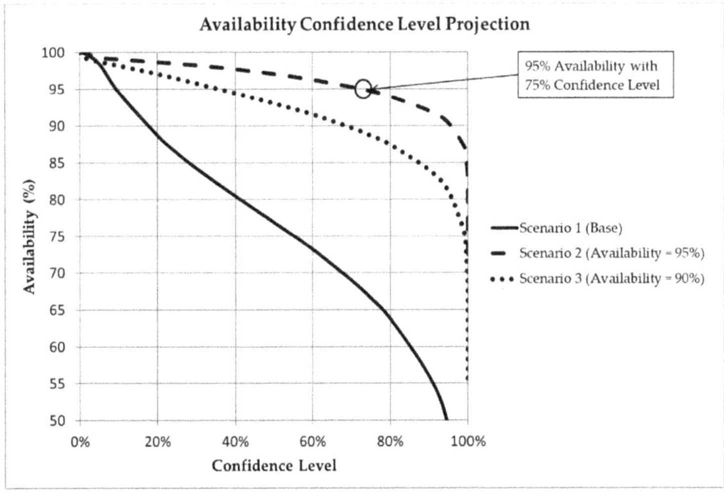

Figure 5. Availability confidence level of the PSS with optimized field kit in place.

In the figure, the topmost curve shows the confidence level of the PSS with the field repair kit optimized for 95% availability. The specific data point singled out shows that this particular PSS with the optimized field repair kit can meet the 95% availability requirement with a 75% confidence level. The second curve shows the confidence level of the PSS with field repair kit optimized for 90% availability. It shows a 70% confidence level for meeting the 90% availability target. Finally, the two optimized field repair kits, in contrast with the original field repair kit in the base scenario, performed particularly well in terms of expected confidence level to meet the PSS availability requirements. Results obtained and shown in Figures 4 and 5 and Table 5 can be a valuable guideline for assessing the PSS availability with a specific field repair kit. Cost savings realized by the adjustment of the field repair kit inventory should be balanced against the expected PSS confidence level for meeting availability requirement promised to customers. The company can also create a mitigation plan based on the confidence level to minimize the risk of not meeting the availability requirement.

4. Conclusions

In this paper, an industrial PSS field repair kit optimization case study of was presented. The objective of the case study was to cost optimize the composition of a field repair kit placed at a customer's site, while maintaining a specific level of PSS availability. A high fidelity PSS simulation model was created and utilized to construct the PSS availability regression model. The availability model was used with other spare parts information to cost optimize the field repair kit, subject to the PSS availability requirements. Monte Carlo simulation was performed to assess the confidence level at which the optimized PSS field repair kit can meet the availability requirements.

Results showed that the PSS field repair kit can be cost optimized to satisfy the availability requirements imposed by the contract with the customer. It was also revealed that the marginal cost of availability improvement was very high, resulting in an almost twofold increase in cost to improve the availability by 5%. The investigation of the optimized PSS repair kit composition showed that the PSS should never shut down for a low cost spare part, but should only shut down for a very expensive spare part that is too expensive to be included in the field repair kit. Finally, Monte Carlo simulation results showed that the optimized PSS field repair kit performed better than the originally proposed field repair kit, meeting the availability requirements at a higher level of confidence. The overall case study results were encouraging in that the modeling and optimization process shown in this paper is applicable to other complex systems.

Future works regarding the improvement of PSS can expand into multiple directions. The latest advances in big data, information and communications technology (ICT), and internet of things (IoT) can greatly improve preventive and predictive maintenance practice for PSS. This can be accomplished through better prediction of parts failure interval and failure modes through analysis of historical data and real-time feedback from the PSS. Big data and IoT can be utilized to design the improved spare parts supply chain as well, providing information for an appropriate inventory level, stocking locations, and reorder points based on PSS location and usage. Another direction the research can take is the improvement of the PSS design: how can PSS be architected for optimum preventive or predictive maintenance policy? How does the PSS architecture impacts spare parts logistics or its supply chain structure? In addition to the topics mentioned, another fundamental direction that the PSS related research can take is to explore ways to reduce the limitation imposed for implementing PSS, such as willingness to adopt the PSS by companies, willingness to accept the PSS by customers, and dealing with environmental implications. All these are very promising future research topics to explore.

Funding: This research received no external funding.

Acknowledgments: The work presented in this paper in its entirety was conducted while employed at Xerox Corporation.

Conflicts of Interest: The author declares no conflict of interest.

References

1. Beuren, F.H.; Ferreira, M.G.G.; Miguel, P.A.C. Product service systems: A literature review on integrated products and services. *J. Clean Prod.* **2013**, *47*, 222–231. [CrossRef]
2. Maussang, N.; Zwolinski, P.; Brissaud, D. Product service system design methodology: From the PSS architecture design to the products specifications. *J. Eng. Design.* **2009**, *20*, 349–366. [CrossRef]
3. Kim, S.H.; Cohen, M.A.; Netessine, S. Performance contracting in after-sales service supply chains. *Manag. Sci.* **2007**, *53*, 1843–1858. [CrossRef]
4. Goedkoop, M.; Van Halen, C.J.; Te Riele, H.; Rommens, P.J. *Product Service Systems, Ecological And Economic Basics*; Report No. 1999/36; Dutch Ministries of environment (VROM) and economic affairs (EZ): The Hague, The Netherlands, 1999.
5. Baines, T.S.; Lightfoot, H.W.; Evans, S.; Neely, A.; Greenough, R.; Peppard, J.; Roy, R.; Shehab, E.; Braganza, A.; Tiwari, A.; et al. State-of-the-art in product service systems. *Proc. Inst. Mech. Eng. B J. Eng. Manuf.* **2007**, *221*, 1543–1552. [CrossRef]
6. Tukker, A. Eight types of product–service system: Eight ways to sustainability? Experiences from SusProNet. *Bus. Strateg. Environ.* **2004**, *13*, 246–260. [CrossRef]
7. Vandermerwe, S. How increasing value to customers improves business results. *Sloan Manag. Rev.* **2000**, *42*, 27.
8. Chiu, M.-C.; Kuo, M.-Y.; Kuo, T.-C. A systematic methodology to develop business model of a product service system. *Int. J. Ind. Eng.* **2015**, *22*, 369–381.
9. Vezzoli, C.; Kohtala, C.; Srinivasan, A. *Product Service System Design for Sustainability*; Greenleaf Publishing: Sheffield, UK, 2014.
10. Rothenberg, S. Sustainability through servicizing. *MIT Sloan Manag. Rev.* **2007**, *48*, 83.
11. Adhikary, D.D.; Bose, G.K.; Jana, D.K.; Bose, D.; Mitra, S. Availability and cost-centered preventive maintenance scheduling of continuous operating series systems using multi-objective genetic algorithm: A case study. *Qual. Eng.* **2016**, *28*, 352–357. [CrossRef]
12. Moghaddam, K.S. Multi-objective preventive maintenance and replacement scheduling in a manufacturing system using goal programming. *Int. J. Prod. Econ.* **2013**, *146*, 704–716. [CrossRef]
13. Mokhtar, A.; Hassene, E.; Laggoune, R.; Chateauneuf, A. Imperfect preventive maintenance policy for complex systems based on Bayesian networks. *Qual. Reliab. Eng. Int.* **2016**, *33*, 751–765. [CrossRef]
14. Van Horenbeek, A.; Pintelon, L. A dynamic predictive maintenance policy for complex multi-component systems. *Reliab. Eng. Syst. Safe* **2013**, *120*, 39–50. [CrossRef]
15. Wang, J.; Zhang, L.; Duan, L.; Gao, R.X. A new paradigm of cloud-based predictive maintenance for intelligent manufacturing. *J. Intell. Manuf.* **2017**, *28*, 1125–1137. [CrossRef]
16. Lee, J.; Kao, H.-A.; Yang, S. Service innovation and smart analytics for industry 4.0 and big data environment. *Procedia CIRP* **2014**, *16*, 3–8. [CrossRef]
17. Yao, B.; Zhou, Z.; Xu, W.; Fang, Y.; Shao, L.; Wang, Q.; Liu, A. Service-oriented predictive maintenance for large scale machines based on perception big data. In Proceedings of the ASME 2015 International Manufacturing Science and Engineering Conference, American Society of Mechanical Engineers, Charlotte, NC, USA, 8–12 June 2015; p. V002T004A015.
18. Xie, W.; Liao, H.T.; Jin, T.D. Maximizing system availability through joint decision on component redundancy and spares inventory. *Eur. J. Oper. Res.* **2014**, *237*, 164–176. [CrossRef]
19. Muckstadt, J.A. Model for a Multi-item, multi-echelon, multi-indenture inventory system. *Manag. Sci.* **1973**, *20*, 472–481. [CrossRef]
20. Sherbrooke, C.C. Multiechelon inventory systems with lateral supply. *Nav. Res. Log.* **1992**, *39*, 29–40. [CrossRef]
21. Shtub, A.; Simon, M. Determination of reorder points for spare parts in a 2-echelon inventory system—The case of nonidentical maintenance facilities. *Eur. J. Oper. Res.* **1994**, *73*, 458–464. [CrossRef]
22. Bijvank, M.; Koole, G.; Vis, I.F.A. Optimising a general repair kit problem with a service constraint. *Eur. J. Oper. Res.* **2010**, *204*, 76–85. [CrossRef]
23. Selcuk, B.; Agrali, S. Joint spare parts inventory and reliability decisions under a service constraint. *J. Oper. Res. Soc.* **2013**, *64*, 446–458. [CrossRef]

24. Kader, B.; Sofiene, D.; Nidhal, R.; Walid, E. Joint optimization of preventive maintenance and spare parts inventory for an optimal production plan with consideration of CO2 emission. *Reliab. Eng. Syst. Saf.* **2016**, *149*, 172–186.

25. Samal, N.K.; Pratihar, D.K. Joint optimization of preventive maintenance and spare parts inventory using genetic algorithms and particle swarm optimization algorithm. *Int. J. Sys. Assur. Eng. Manag.* **2015**, *6*, 248–258. [CrossRef]

26. Cobbaert, K.; VanOudheusden, D. Inventory models for fast moving spare parts subject to "sudden death" obsolescence. *Int. J. Prod. Econ.* **1996**, *44*, 239–248. [CrossRef]

27. Mourtzis, D.; Boli, N.; Alexopoulos, K.; Różycki, D. A framework of energy services: From traditional contracts to Product-Service System (PSS). *Procedia CIRP* **2018**, *69*, 746–751. [CrossRef]

28. Sousa-Zomer, T.T.; Miguel, P.A.C. Sustainable business models as an innovation strategy in the water sector: An empirical investigation of a sustainable product-service system. *J. Clean Prod.* **2018**, *171*, S119–S129. [CrossRef]

29. Charro, A.; Schaefer, D. Cloud manufacturing as a new type of Product-Service System. *Int. J. Comput. Integr. Manuf.* **2018**, *31*, 1018–1033. [CrossRef]

30. Kangz, K.; Zhong, R.Y.; Xu, S. Cloud-enabled sharing in logistics product service system. *Procedia CIRP* **2019**, *83*, 451–455. [CrossRef]

31. Cook, M. Product service system innovation in the smart city. *Int. J. Entrep. Innov.* **2018**, *19*, 46–55. [CrossRef]

32. Tao, F.; Cheng, J.; Qi, Q.; Zhang, M.; Zhang, H.; Sui, F. Digital twin-driven product design, manufacturing and service with big data. *Int. J. Adv. Manuf. Technol.* **2018**, *94*, 3563–3576. [CrossRef]

33. Rahimi-Ghahroodi, S.; Al Hanbali, A.; Vliegen, I.; Cohen, M.A. Joint optimization of spare parts inventory and service engineers staffing with full backlogging. *Int. J. Prod. Econ.* **2019**, *212*, 39–50. [CrossRef]

34. Sleptchenko, A.; Al Hanbali, A.; Zijm, H. Joint planning of service engineers and spare parts. *Eur. J. Oper. Res.* **2018**, *271*, 97–108. [CrossRef]

Article

Research on Optimized Product Image Design Integrated Decision System Based on Kansei Engineering

Lei Xue [1], Xiao Yi [1] and Ye Zhang [2,*]

[1] School of Mechanical, Electronic and Control Engineering, Beijing Jiao Tong University, Beijing 100044, China; 13116347@bjtu.edu.cn (L.X.); xyi@bjtu.edu.cn (X.Y.)

[2] School of Architecture and Design, Beijing Jiao Tong University, Beijing 100044, China

* Correspondence: yzhang4@bjtu.edu.cn

Received: 18 December 2019; Accepted: 7 February 2020; Published: 11 February 2020

Abstract: In order to facilitate the development of product image design, the research proposes the optimized product image design integrated decision system based on Kansei Engineering experiment. The system consists of two sub-models, namely product image design qualitative decision model and quantitative decision model. Firstly, using the product image design qualitative decision model, the influential design elements for the product image are identified based on Quantification Theory Type I. Secondly, the quantitative decision model is utilized to predict the product total image. Grey Relation Analysis (GRA)–Fuzzy logic sub-models of influential design elements are built up separately. After that, utility optimization model is applied to obtain the multi-objective product image. Finally, the product image design integrated decision system is completed to optimize the product image design in the process of product design. A case study of train seat design is given to demonstrate the analysis results. The train seat image design integrated decision system is constructed to determine the product image. This shows the proposed system is effective and for predicting and evaluating the product image. The results provide meaningful improvement for product image design decision.

Keywords: product image design; Kansei Engineering; integrated decision system; qualitative decision model; quantitative decision model; train seats

---

1. Introduction

The development trend of industrial design is toward more efficient, intelligent and systematic. Meanwhile, product design pays more attention to the improvement of the user experience. The development of industrial design decision system not only promotes the economic and cultural exchange between regions, but also has a profound impact on the long-term development of various related industries [1]. With the increasing demand for user experience design, user perception centered design will become an important factor affecting product design [2]. Therefore, based on meeting the product function, we should comprehensively improve the user physiological and psychological multi-dimensional emotional needs, optimize the overall product design level, promote the user experience, and raise the product design quality.

Kansei Engineering is a kind of theory based on the design science, psychology, cognition and other relevant disciplines, which can lead human perceptual analysis to the field of engineering technology [3]. Kansei Engineering not only helps designers understand users' perceptual needs, but also optimizes product design process and reduce product design cost. Kansei Engineering is a method of using engineering technology to explore the relationship between the sensibility of user and the design characteristics of product. It can transform the perceptual needs by users into the design elements of products. In the field of product design, Kansei Engineering is a user-oriented product research and development technology. The modern tools and technologies are used to quantify the user's perceptual information into the design parameters of product, as shown in Figure 1.

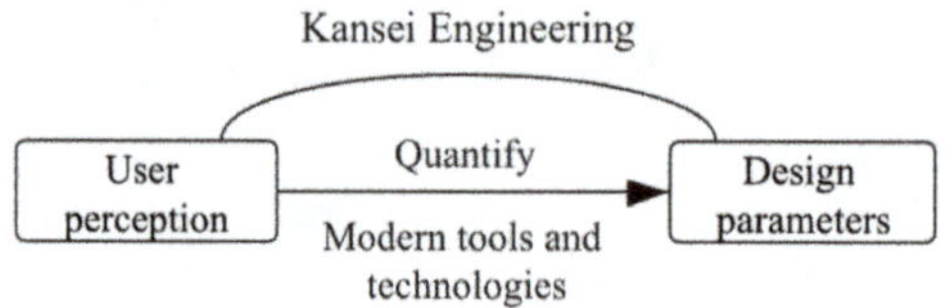

Figure 1. The Kansei Engineering of product design process.

Image is a kind of the psychological symbol that the products give to the users, and is also the psychological image or concept obtained by association and imagination. The users' cognition for the product, through the perception of five senses, causes users' resonance. Finally, a comprehensive psychological image is formed. Image has certain subjectivity and fuzziness. The product appearance design should not only meet the users' basic aesthetic needs, but also meet the users' perception needs. The basic design elements of product constitute the product appearance design. The design elements of product appearance design generally contain form, color, material, texture, pattern and so on. The formation of product image comes from the users' cognition. The design elements of product are taken as the language of communication with the user. Therefore, the formation process of product image is shown in Figure 2.

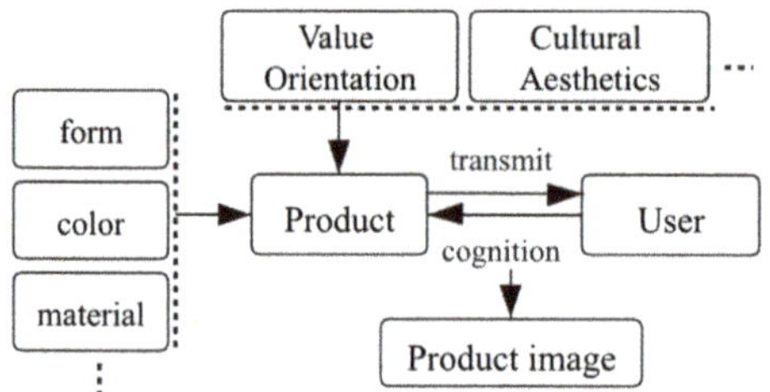

Figure 2. The formation process of product image.

Wei proposed a hybrid adaptive ant colony algorithm to realize product family multi-objective optimization design through scale-based product platform theory model [4]. Lei presented a Decision Support System (DSS) for market-driven product positioning and design, based on market data and design parameters. The proposed DSS determines market segments for new products using Principal Component Analysis, K-means classification [5]. Yang presented a decision support system based on a bi-level fuzzy computing approach that incorporates qualitative and quantitative product attributes in determining the manufacturability of a product [6]. A product design elements evaluation model was proposed, which was constructed by eye-tracking experiments, using eye movement indicators such as first gaze time, gaze order, and number of times of return, to accurately and effectively analyze product model and quantify users' emotion. The purpose of sorting the product model contribution was achieved through the weight calculation of the design elements [7]. Understanding the affective needs of customers is crucial to the success of product design. Hybrid Kansei Engineering system (HKES) is an expert system capable of generating products in accordance with the affective responses. HKES consists of two subsystems: forward Kansei Engineering system and backward Kansei Engineering system [8]. Wu proposed a preference-based evaluation-fuzzy-quantification method to determine the priority of the development of attractive factors of the product [9].

The state of the art has the following deficiencies: (1) The previous research lacks the product image design qualitative decision by the quantitative research methods; (2) It lacks the integrated decision system of product image design for the purpose of the total image design for assistance of the collaborative work of user, designer and expert; (3) The existing product image design model lacks the discussion of the assessment and verification with the computer-aided and 3-D printing of product design.

In our paper, we extend the research objectives to the product material design and texture design. This research proposes a new method for combining the product image design qualitative decision model and quantitative decision model. With the two models, the optimized product image design integrated decision system is constructed based on the Kansei Engineering experiment. In the process of the performance evaluation, the technology of the computer-aided and 3-D printing assist the evaluation with Kansei image of product image design integrated decision system.

Therefore, this paper explores the product image design system under the research framework of Kansei Engineering. The qualitative and quantitative research on the correlation between product design elements and user Kansei image is conducted, and the qualitative and quantitative decision models of product image design are constructed. Then, the integrated decision system of product image design is improved. The paper introduces the methodology that is divided into four steps, shown in Figure 3, including the establishment of qualitative decision model of product image design, quantitative decision model of product image design, product image design integrated decision system and its performance evaluation. Meanwhile, this paper takes the high-speed train seat design as the research object. Kansei Engineering experiment and design theory research are carried out deeply, and the more scientific, efficient and intelligent product image design evaluation system is put forward.

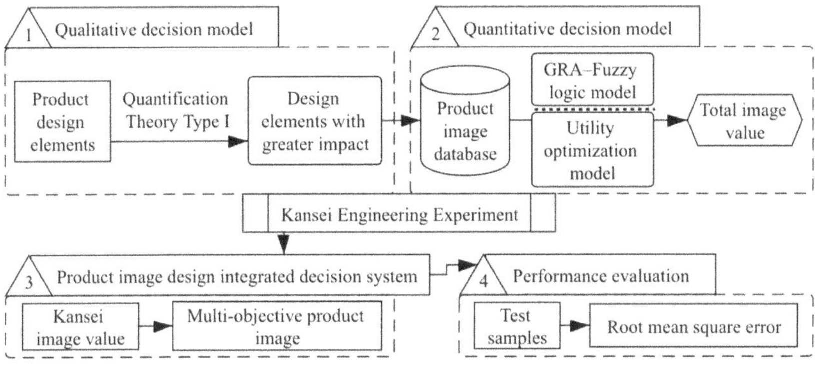

Figure 3. The four steps of the methodology process.

2. Kansei Engineering Experiment

The Kansei Engineering experiment process of product image design integrated decision system mainly includes design investigation stage and data statistics stage as shown in Figure 4. The design investigation stage is divided into two steps. The first step is to select the experimental subjects, experimental samples, product form samples, color samples, material samples and texture samples. Then the second step is to analyze the design elements of experimental design samples separately, such as the morphological analysis pointed to form design element.

The data statistics stage is also divided into two steps. Firstly, the Kansei word pairs are selected, including the selection of the primary Kansei words and secondary Kansei words that describe the product image. Secondly, the Kansei evaluation of the product image is carried out by the subjects for the experimental samples, and the construction of the product image database is built up through the data statistics. Based on the Kansei Engineering process, the qualitative and quantitative decision models of product image design are established to improve the product image design integrated decision system.

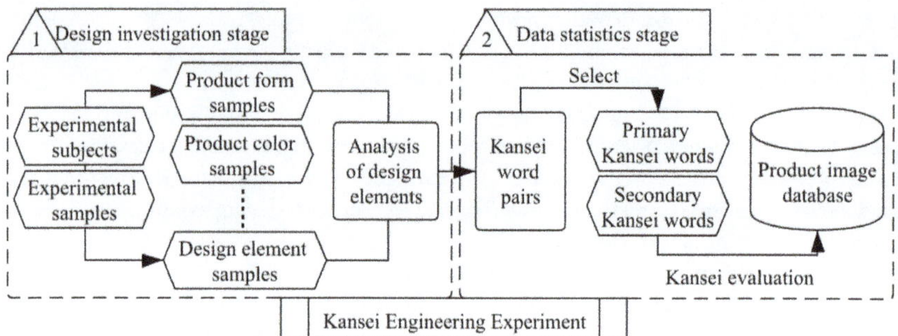

Figure 4. The Kansei Engineering experiment process of product image design.

3. Product Image Design Integrated Decision System

This section may be divided into product image design qualitative decision model, quantitative decision model, integrated decision system and performance evaluation. Firstly, using the Product image design qualitative decision model, the influential design elements for the product image are identified based on Quantification Theory Type I. Secondly, the quantitative decision model is utilized to predict the product total image. GRA–Fuzzy logic sub-models of influential design elements are built up separately. After that, utility optimization model is applied to obtain the multi-objective product image. Thirdly, the product image design integrated decision system is completed to optimize the product image design in the process of product design. Finally, the performance of the system is evaluated by using the root mean square error.

3.1. Product Image Design Qualitative Decision Model

3.1.1. Quantification Theory Type I

The qualitative decision model of product image design is to identify the main design elements affecting product image by quantitative analysis method. The experimental subjects in Section 2 are divided into four groups. The third group is composed of the professional product designers with rich experience in product design. Through the investigation and statistics of professional designers, the four design elements (product form, color, material and texture) are selected for further research of qualitative decision model.

Quantification Theory Type I is the method that studies the relationship between a set of qualitative variables x (independent variables) and a set of quantitative variables y (dependent variables) [10]. The mathematical model between variables is built up by multiple regression analysis. According to the mathematical model of Quantification Theory Type I, the procedure for establishing the Kansei-associated model is as follows:

- The mathematical formula of the Kansei-associated model is defined. The product design element is taken as the independent variable x (qualitative variable) and the Kansei evaluation value is taken as the dependent variable y (quantitative variable). The purpose of Quantification Theory Type I is to solve the regression coefficient of each design element [11]. Therefore, the multiple regression equation of the product design element and Kansei evaluation value can be defined as:

$$\hat{y}_s^k = \sum_{i=1}^{D} \sum_{j=1}^{M_i} \beta_{ij} x_{ijs} \tag{1}$$

where k represents the kth Kansei image, $k = 1, 2, \ldots, m$, and m represents the number of Kansei images; s represents the sth experimental sample, $s = 1, 2, \ldots, n$, and n represents the number of experimental samples; i represents the ith design element, $i = 1, 2, \ldots, D$, and D represents the number of design elements; j represents the jth category of the ith design element, $j = 1, 2, \ldots, M_i$, and M_i represents the number of category for the ith design element.

$\hat{y}_s^k$ is the evaluation value of the kth Kansei image for the sth experimental sample. x_{ijs} represents the independent variable of the jth category of the ith design element for the sth experimental sample, so x_{ijs} can be defined as follows:

$$x_{ijs} = \begin{cases} 1 & \textit{the jth category of the ith design element for the sth experimental sample} \\ 0 & \textit{otherwise} \end{cases} \tag{2}$$

And x_{ijs} meets the condition of $\sum\limits_{j=1}^{M_i} x_{ijs} = 1, \textit{for } \forall i \textit{ and } s.$

β_{ij} represents the partial regression coefficient of the jth category of the ith design element.

- Using the least-squares method, error equation can be defined as below:

$$Q = \sum_{s=1}^{n} \left(y_s^k - \hat{y}_s^k \right)^2 \tag{3}$$

where Q is the sum of square error between the predicted value and the actual value. In order to improve the prediction accuracy and minimize the error, the following equation can be defined for the calculation of partial differential:

$$\frac{\partial Q}{\partial \beta_{ij}} = 0, \textit{ for } \forall \beta_{ij} \tag{4}$$

β_{ij} can be obtained by the simultaneous equations. Namely, the partial regression coefficient of design element can be acquired. The influence degree of each design element on the Kansei image can be obtained by analyzing the numerical value of β_{ij}.

- The complex correlation coefficient R can measure the accuracy of the model. Generally, the square of the complex correlation coefficient R, namely, the coefficient of determination R^2 is used to represent the accuracy of the model. The coefficient of determination is also known as the contribution rate, which is used to measure the cooperation degree or interpretability of the regression equation.
- In order to measure the contribution of each product design element to the prediction evaluation of Kansei image, the partial correlation coefficient needs to be obtained. The partial correlation coefficient indicates the contribution of the ith product design element to the value y of Kansei evaluation. The larger the value of partial correlation coefficient is, the greater the influence of product design element on Kansei image is.

Using Quantification Theory Type I, the partial correlation coefficient of the product design element affecting the Kansei image is obtained, and then the decision coefficient of each design element category is acquired. The partial correlation coefficient represents the contribution of product design element to each Kansei image, and decision coefficient demonstrates the precision of model. In the process of product design, if the designers expect to get the higher evaluation value of specific product image, they should give priority to the product design elements with the higher partial correlation coefficient, properly ignore the design elements with the lower partial correlation coefficient, improve design efficiency, save design cost and optimize product image design process.

3.1.2. Product Image Design Qualitative Decision Model

As shown in Figure 5, the qualitative decision model of product image design is constructed. Firstly, the product design elements are analyzed, including product form, product color, product material, product texture, product pattern and other design elements. The partial correlation coefficient of product element is obtained through Quantification Theory Type I, and the priority ranking of product design elements affecting Kansei image is obtained.

The design elements for the product Kansei image with the greater impact and less impact are analyzed respectively. Designers should pay more attention to the design elements with greater impact, ignore the design elements with less impact, and build a qualitative decision model of product image design. Then, the design elements with greater influence for the profound study are selected, which provides a theoretical basis for the next product image design quantitative decision model.

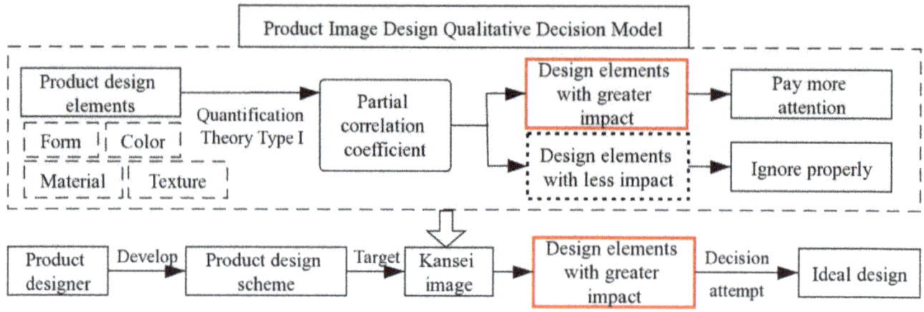

Figure 5. The qualitative decision model of product image design.

3.2. Product Image Design Quantitative Decision Model

3.2.1. GRA–Fuzzy Logic Model

Grey Relation Analysis (GRA) is a method to determine the relationship (similarity) between two sets of random sequences in grey system. One is a reference sequence ($x_0 \in X$) and the other is a comparison sequence $X = \{x_\sigma | \sigma = 0, 1, 2, \ldots, n\}$, and $X = \{x_\sigma | \sigma = 0, 1, 2, \ldots, n\}$ is a set of grey related elements [12]. At a certain moment, the grey correlation degree of two sets of series can be expressed by grey correlation coefficient $r(x_0(k), x_i(k))$, which is defined as below:

$$r(x_0(k), x_i(k)) = \min_i\min_k\left|x_0(k) - x_i(k)\right| + \xi\max_i\max_k\left|x_0(k) - x_i(k)\right|/ \\ \left|x_0(k) - x_i(k)\right| + \xi\max_i\max_k\left|x_0(k) - x_i(k)\right| \tag{5}$$

$k = 1, 2, \ldots, n; i = 1, 2, \ldots, m$

The grey correlation coefficients of different design sub-elements are different. If $r(x_0, x_i) > r(x_0, x_j)$, then design sub-element x_i is more relevant to product image than design sub-element x_j. Based on the Kansei image database of each design sub-element affecting the product image, the grey correlation coefficient is used to express the relationship between the product image and design sub-elements. By comparing the numerical value of grey correlation coefficient, the sorting of the influence degree of different design sub-elements on product image is obtained. The most influential design sub-elements are selected as input linguistic variables, and the corresponding product image as output linguistic variable of GRA–Fuzzy logic model.

Based on the GRA–Fuzzy theory, combining with the Kansei Engineering experiment, the product design sub-elements with the greater influence are selected first, and the product image of the experimental samples is evaluated by seven-point Likert scale [13]. Then, the influential design sub-elements are taken as the input fuzzy set, and the evaluation values of product image are taken as the output fuzzy set. The triangular fuzzy functions of input and output linguistic variables are obtained. Through the application of fuzzy logic model in MATLAB, the fuzzy rules are constructed, and the center of maximum (CoM) method is used for defuzzification [14]. Finally, the GRA–Fuzzy logic model of product image design is established to determine the specific Kansei image value of product design.

Combined with the product image database in Section 2, the fuzzy rules of GRA–Fuzzy logic model are constructed. In order to reflect the relationship between the multiple design sub-elements and product images involved in the product design process, the "if … then … " rules with multiple fuzzy conditions are used as follows:

$$\begin{aligned} &If\ X_1\ is\ A_1\ and\ X_2\ is\ A_2 \cdots and\ X_n\ is\ A_n; \\ &Then\ Y_1\ is\ B_1\ and\ Y_2\ is\ B_2 \cdots and\ Y_n\ is\ B_n \end{aligned} \tag{6}$$

where $A_1, A_2, \ldots, A_n$ and $B_1, B_2, \ldots, B_n$ are fuzzy linguistic terms, which are, respectively, represented by input linguistic variables $X_1, X_2, \ldots, X_n$ and output linguistic variables $Y_1, Y_2, \ldots, Y_n$. The input linguistic variables $X_1, X_2, \ldots, X_n$ represent sub-elements of product design, and the output linguistic variables $Y_1, Y_2, \ldots, Y_n$ represent product image. Each fuzzy rule of GRA–Fuzzy logic model of product image design is to associate the given combination of product design sub-elements with the corresponding numerical state of product image.

Defuzzification is the process of transforming the membership of output linguistic variables into numerical value. The most commonly used defuzzification technology, the center of maximum (CoM), is determined as follows:

$$yCoM = \frac{\sum_i [\mu(y_i \times y_i)]}{\sum_i \mu(y_i)} \tag{7}$$

where i represents the linguistic item of output linguistic variable, y_i is the maximum value of each linguistic item i, and $\mu(y_i)$ is the aggregate output membership function.

Finally, the GRA–Fuzzy logic model of product image design is constructed to assist designers to evaluate the product image design. When the product designers input the combination of product sub-elements, they can output the value of product image and obtain the numerical states of multiple product images.

3.2.2. Utility Optimization Model

In the product design process, it is difficult to judge design scheme by using a single index. For the solution method of multi-objective programming, the utility optimization model is generally used through transforming the multi-objective model into the single-objective model. The model solution basic idea is that each objective function of the programming problems can be calculated in a certain way [15].

$$\max Z = \psi(X) \tag{8}$$

$$s.t.\ \Phi(X) \leq G \tag{9}$$

ψ is the sum function of utility functions related to each objective function.

When using utility function as planning objective, it is necessary to determine a set of w_i that reflects the weight of each objective function in the overall objective of the original problem, namely:

$$\max\psi = \sum_{i=1}^{k} w_i\psi_i \tag{10}$$

$$\Phi(x_1, x_2, \ldots, x_n) \leq g_i (i = 1, 2, \ldots, m) \tag{11}$$

where w_i should satisfy $\sum_i^k w_i = 1$

Parametric vector form is shown as below:

$$\max\psi = w^T\psi \tag{12}$$

$$s.t.\ \Phi(X) \leq G \tag{13}$$

3.2.3. Product Image Design Quantitative Decision Model

The quantitative decision model of product image design is constructed as shown in Figure 6. Based on the Kansei Engineering experiment, the qualitative decision model of product image design in Section 3.1.2 is used to get the influential design elements for product image, and the GRA–Fuzzy logic sub-models of the influential design elements are constructed respectively.

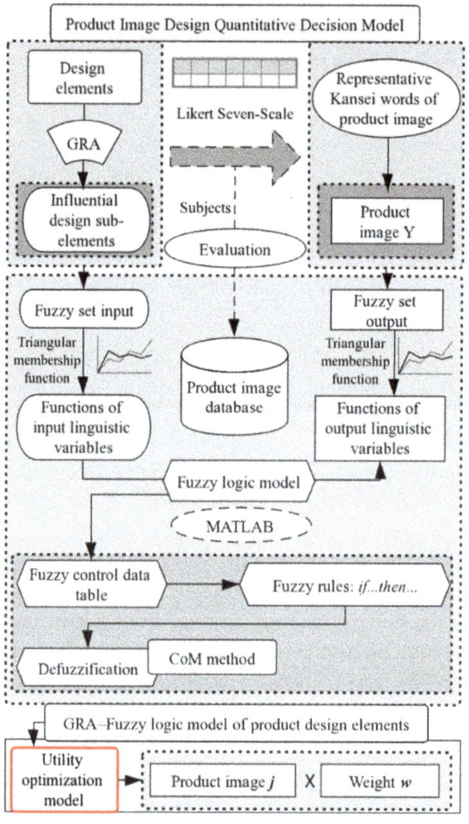

Figure 6. The quantitative decision model of product image design.

Firstly, using the Kansei Engineering experiment process, the design elements with great influence are analyzed, and the sub-elements of the influential design elements are obtained. Meanwhile, the representative Kansei words of the product image expected by users are determined. The subjects evaluate the product image of the experimental samples by Likert scale to set up the product image database. Then, through the grey relational analysis (GRA), the grey correlation coefficients of the design sub-elements that affecting the product image are obtained, and the numerical values of the grey correlation coefficients are sorted and compared in order to obtain the design sub-elements that have a great influence on the product image.

The design sub-elements with great influence obtained above are taken as input linguistic variables, and the product image expected by users is taken as an output linguistic variable. The fuzzy functions of input linguistic variables and output linguistic variables are constructed by the function of triangular membership, and the GRA–Fuzzy logic model is constructed by MATLAB. The fuzzy control data table and the corresponding fuzzy rules between design sub-elements and product image are obtained. Then, the fuzzy numbers are defuzzified by using the CoM method. Finally, the GRA–Fuzzy logic sub-models of the influential design elements are established.

Users usually expect to obtain multiple product images in the practical design and production process. The utility optimization model is used to transform the multi-objective product image into the single-objective product image, optimize the GRA–Fuzzy logic model of product design elements. The weights of product image are obtained according to the demand of design objectives, which reflects the weight of each function of product image in total product image objective, and the product total image value is obtained by linear weighting method. The quantitative decision model of product image design is optimized, namely:

$$g(x) = \sum_{j=1}^{q} w_j g_j(x) \tag{14}$$

where $g_j(x)$ is the GRA–Fuzzy logic model of product image j. w_j is the weight of the jth product image function $g_j(x)$ obtained according to the requirements of objective programming in the product total image objective.

The GRA–Fuzzy logic sub-models of product design elements can help designers evaluate the design elements of product image. Based on the construction of GRA–Fuzzy logic model and utility optimization model, the product image design model is further optimized, and the quantitative decision model of product image design is constructed.

3.3. Product Image Design Integrated Decision System

As shown in Figure 7, product image design integrated decision system is composed of qualitative decision model and quantitative decision model. For the varied kinds of industrial product design, firstly, the qualitative decision model of product image design is constructed to obtain the product design elements that have a greater influence on the user's Kansei image. After that, the quantitative decision model of product image design is implemented to acquire the relationship between the influential design sub-elements and the numerical value of Kansei image. The collaborative work of the above two models constitutes the integrated decision system of product image design. Therefore, it is more efficient and accurate that the combination of product design elements matches with the output of user's Kansei image.

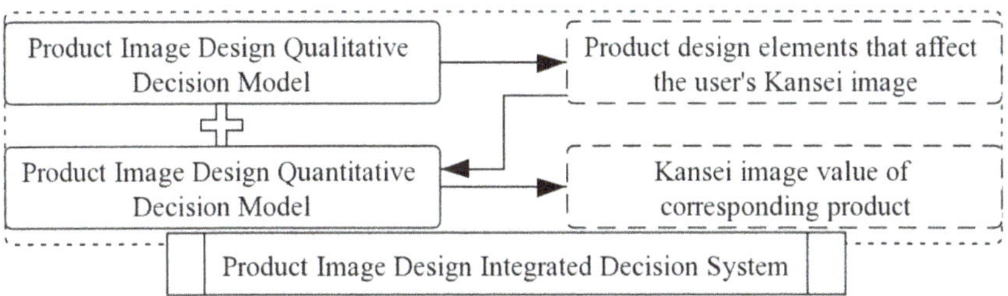

Figure 7. The product image design integrated decision system.

3.4. Performance Evaluation

In order to evaluate the performance of the product image design integrated decision system, the root mean square error (*RMSE*) is used as follows [16]:

$$RMSE = \sqrt{\frac{\sum_{i=1}^{n} (x_i - x_0)^2}{n}} \tag{15}$$

where x_i is the i th output value of product image design integrated decision system and x_0 is the expected value evaluated by the experimental subjects. We use test samples to evaluate the prediction performance of product image design integrated decision system, and the subjects of the second group evaluate the product total image of test samples to get the numerical value. Then, the design sub-elements of each sample are input into the product image design integrated decision system, and the output value of the product total image is obtained, which compared with the evaluation value of the subjects. The *RMSE* of product image design integrated decision system is obtained. Then, the performance evaluation of product image design integrated decision system is completed.

The case study shows that the optimized product image design integrated decision system has a good performance for predicting and evaluating the product image. The proposed system has a better prediction ability of Kansei image for the new product with given design elements.

4. Case Study of Train Seat Image Design

Based on the research of the optimization process of the product image design integrated decision system in Section 3, this section aims at optimization of the high-speed train seat image design decision system. The qualitative and quantitative decision models of the high-speed train seat image design are constructed to further promote the process of the train seat image design integrated decision system.

4.1. Kansei Engineering Experiment

4.1.1. Design Investigation Stage

Firstly, the experimental subjects and samples are selected. Afterwards, the final experimental samples are determined by the selection of train seat form, color, material and texture. Finally, the representative Kansei image words for describing the train seat image are chosen.

- Experimental subject selection

The experimental study involves 60 subjects, divided into four groups. The first group includes general users (10 males and 10 females). In the second group, there are expert users (10 males and 10 females). The third group is composed of 10 professional product designers (6 males and 4 females), and the designers of this group have at least 6 years of product design experience on average. The fourth group includes 10 experts (5 males and 5 females) in the field of high-speed train.

- Experimental sample selection

The subjects of the first group classify the selected product pictures based on their similarity degree of appearance design of train seat, using the Kawakita Jiro (KJ) method. Therefore, the 20 subjects of the first group classify 198 seat pictures of different manufacturers and models that entered the market from 2009 to 2019. Based on the classification results of the subjects, 30 representative experimental samples of train seats are selected by multi-dimensional scale analysis and cluster analysis.

- Design element analysis

The train seat form samples are extracted from 30 representative seat samples. According to the professional knowledge and design experience, 10 designers of the third group obtain the four design elements (including the seat form, color, material and texture) for future analysis. After that, they are required to extract seat form samples through similarity classification, and then 10 subjects are organized into focus groups to summarize and integrate their research results. After that, the 6 form samples are obtained, which are divided into three groups on average, the first group is the form experimental samples. The form experimental sample of train seat No. 2 of the first group is shown in Figure 8.

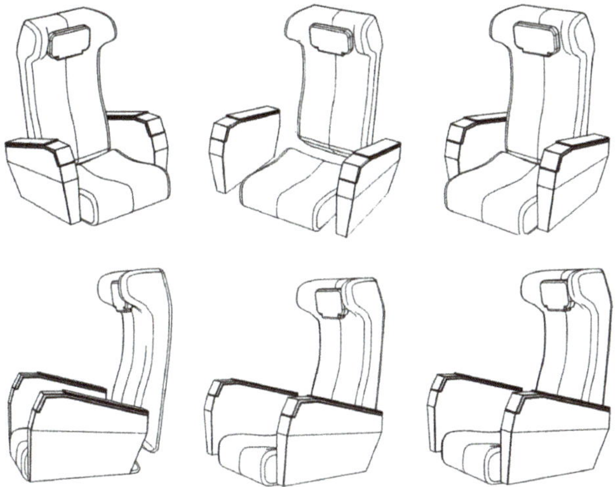

Figure 8. The form experimental sample of high-speed train No. 2 of the first group.

This study regards a small amount of color as a huge work of color speculation and establishes the relationship between product form, color, material, texture and product image [17]. The 10 subjects are organized into a focus group to conduct the inductive integration and analysis of color selection. Finally, the 6 color samples are selected as the color experimental samples of train seat. The form samples are filled with the 6 color experimental samples respectively. The form sample No. 2 is filled with the 6 color experimental samples in Figure 9 as below.

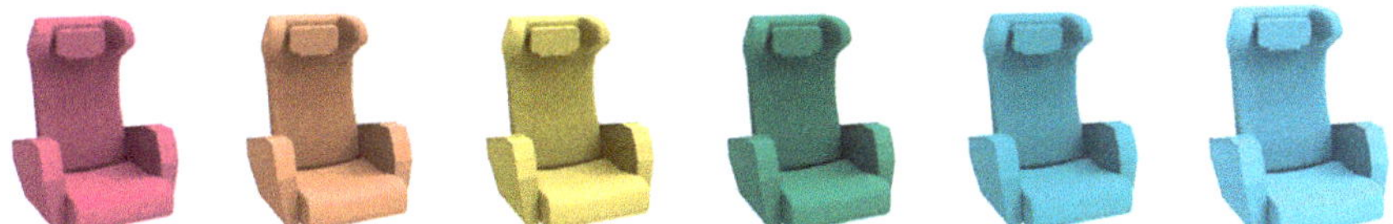

Figure 9. The 6 color experimental samples of the train seat form sample No. 2.

The subjects of third group analyzed the material design and texture design of the train seat, and the representative material samples and texture samples of train seats are collected and selected by the third subjects, and the material samples and texture samples are applied to the color samples. Through the research and investigation, the representative material samples are divided into cotton fabric, wool fabric, chemical fiber and leather, as shown in Table 1.

Table 1. The 4 material experimental samples of the train seat.

Cotton Fabric	Wool Fabric	Chemical Fiber	Leather
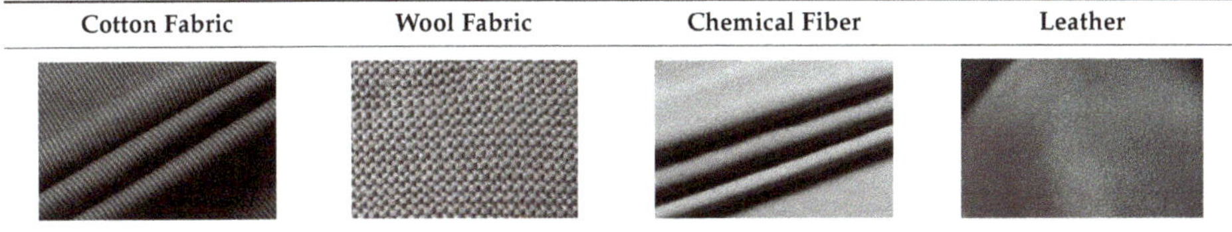			

The representative texture samples are divided into coarse texture and fine texture. The above 4 material samples and 2 texture samples are applied to 12 train seat samples in the first group, a total of 96 train seat experimental samples, as the final experimental samples of this study.

- Kansei image word selection

Based on the preceding steps, three groups of representative Kansei words are determined to describe the Kansei image of train seat: traditional–modern, simple–complex, unique–ordinary. We use "traditional–modern" (T–M) as the primary Kansei word of the train seat, which is the target image for constructing the qualitative decision model and quantitative decision model of the seat image design. The above experimental process becomes the application case for developing the method of Kansei Engineering.

4.1.2. Data Statistics Stage

The expert users evaluate the product image matched with the seat samples. As shown in Figure 10, semantic difference method is used to make the seven-point Likert scale, and 20 subjects are required to carry out the numerical value evaluation of T–M product image for the experimental samples (1 and 7 represent the extremely traditional and extremely modern image respectively). For example, seat No. 2 has "modern" image, namely that the maximum value of T–M image is 7, the minimum value is 3, and the average value is 5.33. The data statistics stage provides the experimental database for the qualitative decision model and quantitative decision model of seat image design.

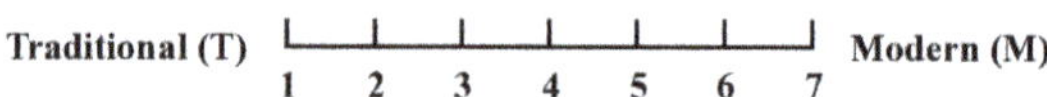

Figure 10. The seven-point Likert scale of T–M product image.

4.2. Train Seat Image Design Qualitative Decision Model

Through the analysis of Quantification Theory Type I, the importance and influence degree of train seat design elements are obtained. The independent variables (such as seat form, color, material and texture) of factors quantification are the nominal scales, so the numerical values of 1, 2, 3 and 4 are used to represent the categories of design elements, for example, "cotton fabric", "wool fabric", "chemical fiber" and "leather" of material samples are, respectively, represented by 1, 2, 3 and 4. The result is shown in Table 2.

Table 2. Analysis of result from Quantification Theory Type I.

Design Element	Standard Coefficient	Partial Correlation Coefficient
Form element (X_1)	0.290	0.441
Color element (X_2)	0.769	0.791
Material element (X_3)	−0.205	−0.290
Texture element (X_4)	0.002	0.003
Constant	2.464	
Coefficient of determination R^2	0.655	

The partial correlation coefficients of design elements in Table 2 show that the correlation between four independent variables (seat form, color, material and texture) and Kansei image words (T–M). For example, for the T–M image, the partial correlation coefficients of design elements are 0.441, 0.791, 0.290 and 0.003, respectively, X_2 of which is the highest, indicating that the "color" factor is the design factor that has the greatest influence on T–M image of the independent variables. The second factor is seat form (X_1), which shows that when the design expectation of the train seat is to obtain the T–M image, the designer should pay more attention to color design element and form design element. On the contrary, designers can properly ignore the design elements with less relevance, such as seat material (X_3) and texture (X_4), because these two design elements have a relatively small impact on T–M image.

Based on the above experimental process of Kansei Engineering, the qualitative decision model of train seat image design is finally obtained by using Quantification Theory Type I method. The qualitative decision model of train seat image design is shown in Figure 11. The partial correlation coefficients of seat design elements are obtained through Quantification Theory Type I. The sequence of design elements affecting the Kansei T–M image of train seat is color, form, material and texture. Therefore, the study selects the design elements that have a great influence on the Kansei image: seat form and seat color, which provides a theoretical basis for constructing the quantitative decision model of train seat image design in the next step.

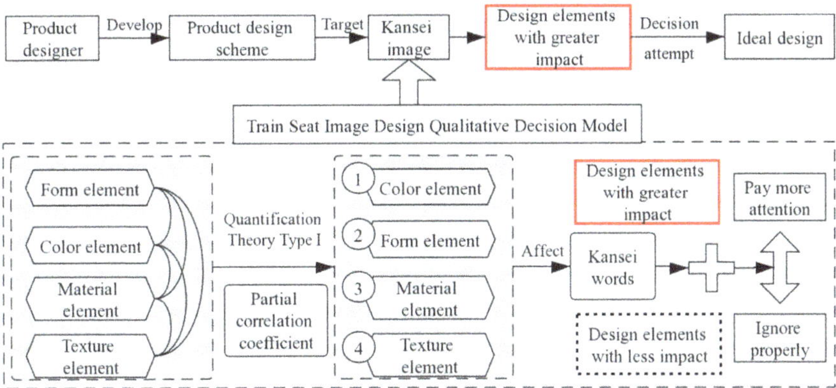

Figure 11. The qualitative decision model of train seat image design.

4.3. Train Seat Image Design Quantitative Decision Model

Based on the conclusion of train seat image design qualitative decision model, the design elements that have great influence on seat image are seat form and color design elements. Using the experimental process of Kansei Engineering, the GRA–Fuzzy logic sub-models of train seat form design and color design are respectively constructed. Based on the GRA–Fuzzy theory, this paper analyzes the form sub-elements and the color sub-elements. Aimed at the representative Kansei words of the seat image, the subjects evaluate the Kansei image of seat samples through the seven-point Likert scale to build up the train seat image database.

As the result of design element analysis, the 9 form sub-elements and their corresponding element types are extracted from the 30 representative train seat samples through morphological analysis. In this section, we select one representative train seat randomly from the 30 representative train seat samples and fill the selected train seat with 72 representative colors as the experimental train seat color samples, for a total of 72 color samples. In Table 3, Column 1 and 7 display the numbers of color samples, and the remaining columns display the corresponding color sub-element ("Hue ($h°$)" element (x_{c1}), "Chroma (C^*)" element (x_{c2}), "Lightness (L^*)" element (x_{c3}), "parameter a^*" element (x_{c4}) and "parameter b^*" element (x_{c5})) values of these color samples.

Table 3. Color element values of the representative train seat color sample C22–C33.

No.	$h°$	C^*	L^*	a^*	b^*	No.	$h°$	C^*	L^*	a^*	b^*
C22	90	60	50	0	60	C28	135	60	30	−42	42
C23	90	40	50	0	40	C29	135	40	30	−28	28
C24	90	20	50	0	20	C30	135	20	30	−14	14
C25	90	60	70	0	60	C31	135	60	50	−42	42
C26	90	40	70	0	40	C32	135	40	50	−28	28
C27	90	20	70	0	20	C33	135	20	50	−14	14

The GRA method is used to obtain the grey correlation degree of form design sub-element as shown in Table 4 from Section 3.2.1, and the sub-elements of form design that affect the Kansei image of train seat are, respectively, "Seat back waistline" (x_5), "Seat back body shape" (x_4), "Headrest shape" (x_2), "Length ratio between seat-back to seat" (x_8), "Relationship between seat-back and headrest" (x_3) and "Seat-back slope angle" (x_9). The above form design sub-elements that have greater influence on Kansei image are taken as the input linguistic variables and the Kansei image of train seat as output linguistic variables. The GRA–Fuzzy logic sub-model between the form design sub-elements and Kansei image is constructed by the triangular membership function.

Table 4. The numerical values of $r(x_0, x_i)$.

Grey Correlation Degree		
$r(x_0, x_1) = 0.722$	$r(x_0, x_2) = 0.812$	$r(x_0, x_3) = 0.794$
$r(x_0, x_4) = 0.833$	$r(x_0, x_5) = 0.844$	$r(x_0, x_6) = 0.515$
$r(x_0, x_7) = 0.772$	$r(x_0, x_8) = 0.798$	$r(x_0, x_9) = 0.787$

Afterwards, the corresponding 54 fuzzy rules are built up, and the GRA–Fuzzy logic sub-model of train seat form design is constructed. The fuzzy rules No. 22–No. 30 have been added as followed in Table 5.

Table 5. The fuzzy rules No.22–No.30 of GRA–Fuzzy logic sub-model of form design sub-elements.

Rule No.	If						Then	
	x_2	x_3	x_4	x_5	x_8	x_9	Y(T–M)	Dos
22	I	J	PL	AL	MR	SA	M	0.75
23	I	J	IC	AL	SR	LA	M	0.58
24	I	J	IC	AL	SR	LA	VM	0.42
25	E	I	RC	IL	MR	LA	VM	0.83
26	E	I	RC	IL	MR	LA	M	0.17
27	E	S	PL	IL	SR	SA	M	0.83
28	E	S	PL	IL	SR	SA	N	0.17
29	S	I	RC	PL	SR	SA	N	0.75
30	S	I	RC	PL	SR	SA	M	0.25

According to Section 3.2.1, GRA method is used to obtain the grey correlation degree of color design sub-element. It shows that the color design sub-elements that affect the Kansei image of train seat are "Lightness (L^*)" (x_{c3}), "Chroma (C^*)" (x_{c2}), "parameter a^*" (x_{c4}) and "parameter b^*" (x_{c5}). Taking the above color design sub-elements that have great influence on seat Kansei image as input linguistic variables and seat Kansei image as output linguistic variables, the GRA–Fuzzy logic sub-model of color design sub-elements matched with Kansei image is constructed by triangular membership function. Next, the 144 corresponding fuzzy rules are established, and the GRA–Fuzzy logic sub-model of train seat color design is constructed.

Based on the cooperative work of GRA–Fuzzy logic sub-models of the aforementioned train seat form design and color design, in terms of the application of utility optimization model, the utility optimization model is constructed for the Kansei image output value of train seat for perceptual cognition of users. Three groups of representative Kansei words are selected, including traditional–modern (T–M), simple–complex (S–C) and unique–ordinary (U–O).

In accordance with the method of multi-image objective programming, three corresponding GRA–Fuzzy logic sub-models of T–M, S–C and U–O images are established, respectively. When using utility function as the planning objective, three corresponding weights w_1, w_2 and w_3 are used to reflect the weight of each GRA–Fuzzy logic sub-model of each Kansei image in the overall objective, and the utility optimization model of the product total image is obtained. When the primary image of seat design expected by users is T–M image, followed by S–C and U–O image, according to the planning of users' perceptual needs, the weight of GRA–Fuzzy logic sub-model for T-M image is 0.4, and the weight of GRA–Fuzzy logic sub-model reflecting S–C image and U–O image is 0.3, respectively. Finally, the seat design total image of multi-image planning objective of train seat is obtained.

Based on the experiment process of Kansei Engineering, combined with the utility optimization model, the seat image design quantitative decision model of high-speed train is finally constructed, and the specific process is shown in Figure 12.

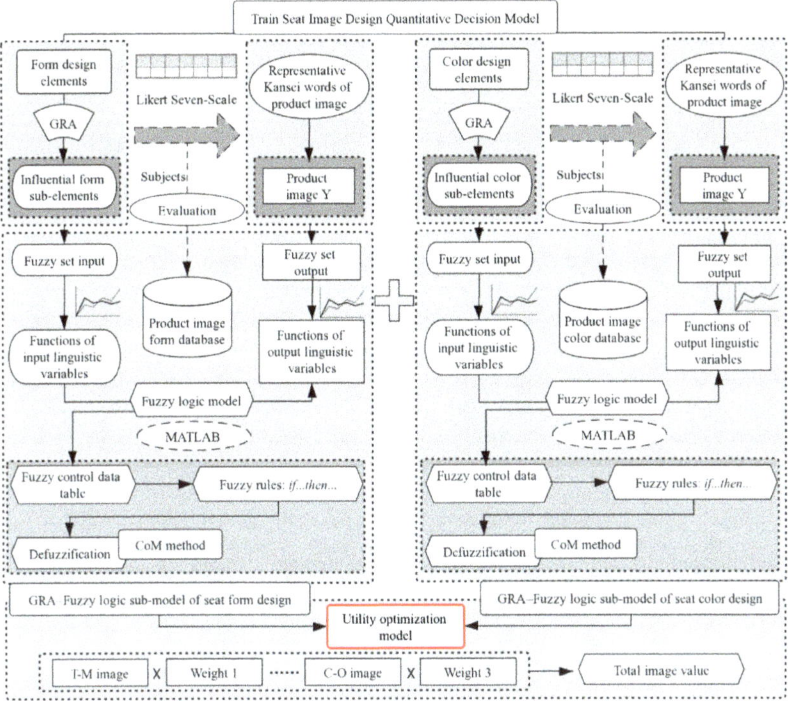

Figure 12. The quantitative decision model of train seat image design.

4.4. Train Seat Image Design Integrated Decision System

As shown in Figure 13, the train seat image design integrated decision system consists of qualitative decision model and quantitative decision model of seat image design. Aiming at the seat image design of high-speed train, the qualitative decision model of the seat image design is constructed firstly, and then the seat form design elements and color design elements that affect the Kansei image are obtained. After that, the quantitative decision model of the seat image design is constructed to obtain the predicted numerical value of the total image of the corresponding seat. The collaborative work of the above two models constitutes the integrated decision system of train seat image design, which makes the image design process of train seat more intelligent, accurate and efficient.

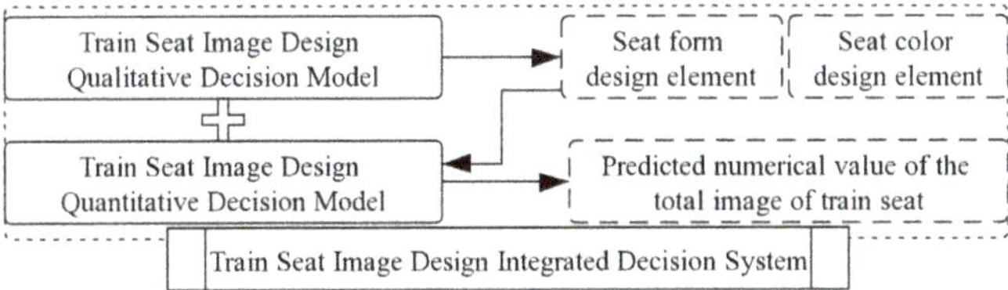

Figure 13. The train seat image design integrated decision system.

4.5. Performance Evaluation

In order to verify the prediction ability of train seat image design integrated decision system, five test samples shown in Table 6 are used, including form design sub-elements and color design sub-elements that have greater influence on product image. The performance of train seat image design integrated decision system is evaluated by using the root mean square error.

Table 6. Input and output values of the seat test samples.

Sample	x_2	x_3	x_4	x_5	x_8	x_9	x_{c2}	x_{c3}	x_{c4}	x_{c5}	Y(T–M)
1	SE	S	RC	CL	MR	SA	20	30	−14	14	3.90
2	I	J	IC	CL	MR	LA	60	70	0	−60	6.35
3	E	S	IC	IL	MR	LA	60	30	−42	42	4.31
4	I	S	IC	IL	MR	LA	20	70	0	20	4.00
5	SE	I	PL	FL	SR	SA	60	50	0	−60	4.73

We select T–M image as the primary image of seat design, then S–C and U–O image as the secondary image, and apply the seat image design integrated decision system to predict the product total image. The second row of Table 7 shows the average value of the product total image evaluated by the 20 subjects of second group on five test samples, and meanwhile, the third row indicates the prediction value of the integrated decision system of seat image design. The results show that the seat image design integrated decision system has lower *RMSE* value. Therefore, the integrated decision system of seat image design provides an effective mechanism for train seat design to match with the total image of product design expected by users.

Table 7. *RMSE* result of the integrated decision system of train seat total image design.

Total Image	Sample 1	Sample 2	Sample 3	Sample 4	Sample 5	*RMSE*
Evaluation by subjects	4.19	5.69	4.22	4.07	4.67	
Integrated decision system	4.40	5.48	4.26	4.34	4.54	0.1895

The good performance of the seat image design integrated decision system shows that the system can help the designer to evaluate the combination of train seat design elements, aiming at satisfying the expected Kansei image. Therefore, this paper provides an effective design evaluation system for the improvement of seat design reflecting the ideal product image through the product image design integrated decision system. For example, the product designer can input the values of multiple design sub-elements by using the product image design integrated decision system, and then the predicted value of multi-objective product image is obtained. If the predicted value does not meet the standard of designer, the corresponding new product image predicted value can be obtained by modifying the combination of design sub-elements until the designer obtains the satisfactory product image value.

4.6. Computer-Aided Seat Design and Image Evaluation

Combined with the computer-aided seat form design and color design, the seat form and color design elements are input into the train seat image design integrated decision system, and then the product image value is obtained. Through the further application of the computer-aided product form modeling and color rendering, train seat image design integrated decision system is verified for the predictable performance of product image.

Based on the computer-aided seat test sample form design and color design, four color design samples are applied to three form design samples, and a total of 12 product design samples are obtained. Figure 14 shows the seat rendering of the form sample 3 filled with the color sample 1.

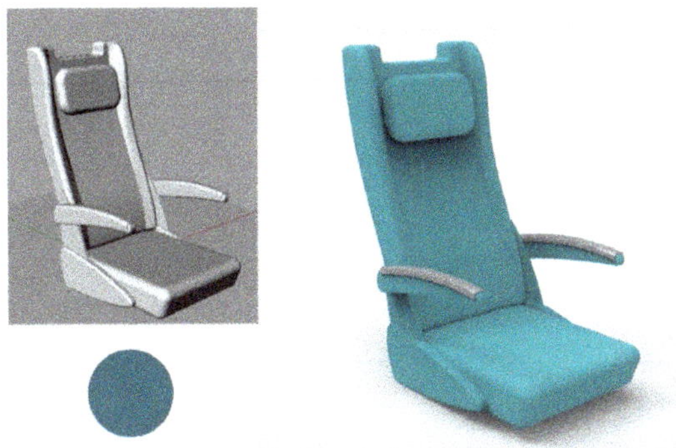

Figure 14. The train seat rendering of the form sample 3 filled with the color sample 1.

At first, the experimental subjects are required to evaluate the T–M image of the above 3-D seat models. After that, the T–M image outputs of the seat models are carried out by using the train seat image design integrated decision system. The T-M image results of models 5 to 10 are shown in the third row of Table 8. Finally, the *RMSE* value of the train seat image design integrated decision system is calculated with the result of less than 0.5, which proves that integrated decision system of train seat image design performs better in predicting the T–M image.

Table 8. *RMSE* result of the integrated decision system of train seat T–M image design.

T–M Image	Model 5	Model 6	Model 7	Model 8	Model 9	Model 10	*RMSE*
Evaluation by subjects	3.86	5.12	4.84	5.92	3.22	5.12	
Integrated decision system	4.03	5.45	5.35	5.69	3.53	4.95	0.3098

In conclusion, the integrated decision system of seat image design can effectively predict the image value of 3-D seat model of computer-aided design and rendering. Therefore, utilizing computer software to build 3-D model of industrial product, combined with computer-aided 3-D printing rapid prototyping, we can assist the use of product image design integrated decision system, which makes the image evaluation of product design more intuitive and efficient. Figure 15 illustrates the 3-D printed drawing of the form sample 3. In this study, 3-D model printing is carried out according to the ratio of 5:1 reduction of seat 3-D model to actual size in order to assist the evaluation process of integrated decision system of seat image design.

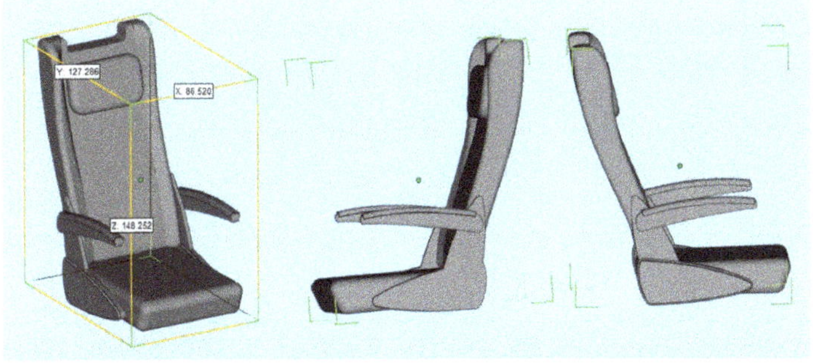

Figure 15. The 3-D printed drawing of the seat form sample 3.

5. Research Limitations and Discussions

This paper takes the seat image design of high-speed train as the research case, and this method is also applicable to other industrial product design, interaction design, service design and user research fields. Based on the experimental framework of Kansei Engineering, the Kansei image perceived by users is transformed into product design elements, interaction design elements or service design elements, etc. Moreover, by establishing quantitative and qualitative decision models of product image design, the integrated decision system of product image design is obtained. The research system can evaluate the perceptual experience of users, make the designers work more efficiently and accurately, and promote and optimize the process of product design, interaction design, service design and user research by predicting the Kansei image of products.

The next work is to further explore the product image design integrated decision system based on Kansei Engineering, combined with product environment elements. The integrated decision system will be expanded and upgraded in the future. With the change of society and the passing of time, the perceptual cognition of users will change constantly. The evaluation database of experimental subjects and the Kansei image database of products in the integrated decision system of product image design need to be adjusted and updated timely to ensure that the integrated decision system of product image design can go with the tide of development and the user environment for any new product market.

In the intelligent application environment, more high and new technologies are developing rapidly. Therefore, the development trend of the product image design integrated decision system will be more intelligent, accurate, digital and systematic. Meanwhile, in the establishment process of experimental database, the design concept, 3-D model and rendering of product design will be further optimized and promoted [18]. Therefore, the intelligent research of product image design integrated decision system has broad prospects and far-reaching research significance.

6. Conclusions

In this paper, we have proposed a methodology of product image design integrated decision system based on Kansei Engineering theory. Firstly, Quantification Theory Type I is used to effectively identify the influential design elements that affect the product image, and the product image design qualitative decision model is constructed. Secondly, according to GRA–Fuzzy theory, the GRA–Fuzzy logic sub-models of influential design elements for the product image are constructed. Then, combining with the utility optimization model, the product image design quantitative decision model is built up. Finally, the product image design integrated decision system is constituted to analyze, decide and evaluate the product image design. The result shows that the system is effective in predicting the product images and optimizes the product image design process.

To illustrate the product image design integrated decision system, we have performed an experimental study on train seat design. The result demonstrates that the system has a better performance for improving the product image design. The designers can effectively plan the product design for the specific product image to meet the user perceptual expectation. Although the train seat design is used as a case in this paper, the methodology can be applied to other user-oriented product designs with multiple design elements. Along with the development of modern science and technology innovation and development, the product image design integrated decision system involved with multiple product design elements and product images can be further improved and developed in the future.

Author Contributions: Conceptualization, L.X. and X.Y.; methodology, L.X.; software, L.X.; validation, L.X. and Y.Z.; formal analysis, L.X.; investigation, Y.Z.; resources, X.Y.; writing—original draft preparation, L.X.; writing—review and editing, L.X., X.Y. and Y.Z.; visualization, L.X. and Y.Z.; supervision, X.Y.; project administration, X.Y.; funding acquisition, L.X. and X.Y. All authors have read and agreed to the published version of the manuscript.

Funding: This research was funded by the Major Program of Beijing Municipal Science and Technology Commission under grant number D16111000110000.

Acknowledgments: We would like to thank the 60 subjects, 10 product designers, 10 design experts for their participation and assistance in the experimental study. We also appreciate the editor and anonymous reviewers for their valuable comments.

Conflicts of Interest: The authors declare no conflict of interest.

References

1. Sarah, L.; John, V. Industrial Design for Modern Life by Danielle Shapiro. *Technol. Cult.* **2018**, *59*, 490–492.
2. Zheng, P.; Yu, S.; Wang, Y.; Zhong, R.Y.; Xu, X. User-experience based product development for mass personalization: A case study. *Procedia CIRP* **2017**, *63*, 2–7. [CrossRef]
3. Li, Y.; Shieh, M.D.; Yang, C.C. A posterior preference articulation approach to Kansei engineering system for product form design. *Res. Eng. Des.* **2019**, *30*, 3–19. [CrossRef]
4. Wei, W.; Tian, Z.; Peng, C.; Liu, A.; Zhang, Z. Product family flexibility design method based on hybrid adaptive ant colony algorithm. *Soft Comput.* **2018**, *23*, 509–520. [CrossRef]
5. Lei, N.; Moon, S.K. A Decision Support System for market-driven product positioning and design. *Decis. Support Syst.* **2015**, *69*, 82–91. [CrossRef]
6. Yang, S.S.; Ong, S.K.; Nee, Y.C. A Decision Support Tool for Product Design for Remanufacturing. *Procedia CIRP* **2016**, *40*, 144–149. [CrossRef]
7. Yang, W.; Su, J.; Qiu, K.; Zhang, X.; Zhang, S. Research on Evaluation of Product Image Design Elements Based on Eye Movement Signal. In Proceedings of the International Conference on Human-Computer Interaction, Paphos, Cyprus, 2 September 2019; Springer: Cham, Switzerland.
8. Shieh, M.D.; Li, Y.; Yang, C.C. Comparison of multi-objective evolutionary algorithms in hybrid Kansei engineering system for product form design. *Adv. Eng. Inform.* **2018**, *36*, 31–42. [CrossRef]
9. Wu, Y.; Cheng, J. Continuous Fuzzy Kano Model and Fuzzy AHP Model for Aesthetic Product Design: Case Study of an Electric Scooter. *Math. Probl. Eng.* **2018**, *2*, 1–13. [CrossRef]
10. Florio, S.; Jones, N.K. Unrestricted Quantification and the Structure of Type Theory. *Philos. Phenomenol. Res.* **2019**, *8*, 1–21. [CrossRef]
11. Chang, H.-C.; Chen, H.-Y. Exploration of Action Figure Appeals Using Evaluation Grid Method and Quantification Theory Type I. *Eurasia J. Math. Sci. Technol. Educ.* **2016**, *13*, 1445–1459. [CrossRef]
12. Das, R.; Ball, A.K.; Roy, S.S. Optimization of E-jet Based Micro Manufacturing Process Using Grey Relation Analysis. *Mater. Today Proc.* **2018**, *5*, 200–206. [CrossRef]
13. Ivanov, O.A.; Ivanova, V.V.; Saltan, A.A. Likert-scale questionnaires as an educational tool in teaching discrete mathematics. *Int. J. Math. Educ.* **2018**, *1*, 1–9. [CrossRef]
14. Rikalovic, A.; Cosic, I.; Labati, R.D.; Piuri, V. Intelligent Decision Support System for Industrial Site Classification: A GIS-Based Hierarchical Neuro-Fuzzy Approach. *IEEE Syst. J.* **2018**, *12*, 2970–2981. [CrossRef]
15. Duffuaa, S.O.; Gaaly, A.E. A multi-objective optimization model for process targeting with inspection errors using 100% inspection. *Int. J. Adv. Manuf. Technol.* **2017**, *88*, 1–14. [CrossRef]
16. Lin, Y.C.; Wei, C.C. A hybrid consumer-oriented model for product affective design: An aspect of visual ergonomics. *Hum. Factors Ergon. Manuf. Serv. Ind.* **2017**, *27*, 17–29. [CrossRef]
17. Xue, L.; Yi, X.; Lin, Y.C.; Drukker, J.W. A method of the product form design and color design of train seats based on GRA-Fuzzy theory. *J. Eng. Technol.* **2018**, *6*, 517–536.
18. Zhang, D.; Li, Z.; Qin, S.; Han, S. Optimization of Vibration Characteristics of Fused Deposition Modeling Color 3D Printer Based on Modal and Power Spectrum Method. *Appl. Sci.* **2019**, *9*, 41–54. [CrossRef]

Article

A Coordination Space Model for Assemblability Analysis and Optimization during Measurement-Assisted Large-Scale Assembly

Zhizhuo Cui and Fuzhou Du *

School of Mechanical Engineering & Automation, Beihang University, Beijing 100191, China; by1407136@buaa.edu.cn
* Correspondence: du_fuzhou@buaa.edu.cn; Tel.: +86-1352-182-3595

Received: 17 April 2020; Accepted: 8 May 2020; Published: 11 May 2020

Featured Application: For assembly incoordination caused by excessive assembly deviations, the proposed method can predict the assemblability and solve the assembly features that need accuracy compensation, to improve the assembly efficiency.

Abstract: The assembly process is sometimes blocked due to excessive dimension deviations during large-scale assembly. It is inefficient to improve the assembly quality by trial assembly, inspection, and accuracy compensation in the case of excessive deviations. Therefore, assemblability prediction by analyzing the measurement data, assembly accuracy requirements, and the pose of parts is an effective way to discover the assembly deviations in advance for measurement-assisted assembly. In this paper, a coordination space model is constructed based on a small displacement torsor and assembly accuracy requirements. An assemblability analysis method is proposed to check whether the assembly can be executed directly. Aiming at the incoordination problem, an assemblability optimization method based on the union coordination space is proposed. Finally, taking the space manipulator assembly as an example, the result shows that the proposed method can improve assemblability with a better assembly quality and less workload compared to the least-squares method.

Keywords: measurement-assisted assembly; coordination space; assemblability; small displacement torsor

1. Introduction

Large-scale mechanical products like ships, automobiles, aircrafts, etc. are complex in structure, large in size, and accurate in assembly quality. The assembly workload of the manufacturing process is heavy [1]. These products often need accuracy compensation in the assembly process because of the excessive assembly deviations, which lead to inefficiency. The assembly deviations might be caused by the eventual poor machining quality of parts, or excessive tolerances set by designers. Thus, the trial assembly is often used to detect the assembly deviations in advance, and the parts are then separated to make an accuracy compensation on the bad dimensions. The assembly process takes a long time by the following steps: Trial assembly, measurement of deviations, separation of parts, and re-trial assembly. Therefore, an assemblability analysis and optimization method based on the measurement data is necessary to predict the assembly deviation and make the accuracy compensation in advance.

With the development of measurement-assisted assembly (MAA) [2], measurement technology has become a bridge between the real world and the digital world. Marguet et al. [3] introduced a MAA application in an airbus assembly line. The least-squares method was used to calculate the optimal pose. Chen et al. [4] proposed a weighted SVD algorithm to obtain the optimal pose of components, which improved the accuracy of pose evaluation. Li et al. [5] proposed a coaxial alignment method using distributed monocular vision. The iterative reweighted particle swarm optimization method was constructed to improve the measurement ability of complicated wearing holes. Wang et al. [6] calculated the assembly clearance of a wing-fuselage assembly based on the optimal pose. The above methods mainly consider the measurement and calculation of the assembly pose, and then realize alignment through pose adjustment tooling. The assembly will be difficult if the quality of the parts is poor.

Assemblability prediction is the first step to judge whether the assembly is qualified in the measurement-assisted assembly. Sukhan et al. [7] evaluated the assemblability based on tolerance propagation. Sanderson et al. [8] assessed the assemblability by the maximum likelihood problem, which was solved by the Kalman filter algorithm. The traditional assemblability evaluation methods are mainly used to find the assembly problem in the design phase, but not in the assembly phase. Cui and Du [9] proposed the concept of pose feasible space to assess the assembly coordination. Yuan et al. [10] proposed an assembly quality assessment method based on weighted geometric constraints to calculate the optimal pose. Wu et al. [11] proposed a constraint coordination index to assess the assembly quality. Ma et al. [12] developed the assembly precision pre-analysis technique in the simulation of virtual assembly. Du et al. [13] proposed a pose decoupling model of the axis tolerance feature to decouple the analysis of any pose within the tolerance domain.

The accuracy compensation methods are used to improve assemblability. The digital compensation method has become a research highlight to improve the assemblability. Davis et al. [14] put forward the method of measuring the assembly clearance and realizing the digital manufacturing of the accuracy compensation gasket. Fabian et al. [15] introduced a shimming method by 3D printing technology, and the assembly clearance was measured by optical measurement. Wang et al. [16] provided a shimming method based on scanned data for a wing box assembly involving non-uniform gaps. In addition, finite element analysis was taken to improve the shimming scheme. Those methods, however, need to be assembled first, followed by measurement of the deviations to be compensated, resulting in a lower efficiency.

Some scholars proposed predictive shimming and predictive fettling methods to improve the assembly efficiency and quality [17]. Cui et al. [18] proposed the oriented points group to calculate the deviation of multiple shaft-and-holes, and the gap was shimmed. Yang et al. [19] analyzed the deviation from the measured point cloud to the model to improve skin finishing. Yu et al. [20] employed a virtual assembly and repair analysis method based on both the geometric design model and object scanning model. Manohar et al. [21] proposed an alternative strategy for predictive shimming, based on machine learning and sparse sensing to first learn gap distributions from historical data. Lei et al. [22] presented an automated and in situ alignment approach with the assistance of computer numerical controlled (CNC) positioners and laser trackers to reduce the finish machining workload. The above studies are aimed at specific cases.

The accuracy compensation method is usually applied after assembly. Then, the assembly sometimes needs be separated, which leads to low efficiency. In this paper, an assemblability analysis and optimization method based on the coordination space model is constructed during measurement-assisted large-scale assembly. In Section 2, the coordination space model based on the small displacement torsor is constructed. In Section 3, the assemblability analysis based on the coordination space model is proposed. In addition, the uncoordinated case is further analyzed. In Section 4, the assemblability optimization method based on the union coordination space is proposed for the uncoordinated case. In Section 5, the space manipulator assembly is taken as an example to verify the proposed method. The result shows that the proposed method can optimize the assemblability with less workload and better assembly quality compared to the least-squares method.

2. Coordination Space Model Based on Small Displacement Torsor

Assemblability refers to the ability of parts to satisfy the assembly accuracy requirements in terms of dimensions, which can be expressed by coordination accuracy. Traditionally, coordination accuracy [23] is the difference in the manufacturing dimensions. Figure 1 shows the coordination accuracy of a keyway assembly.

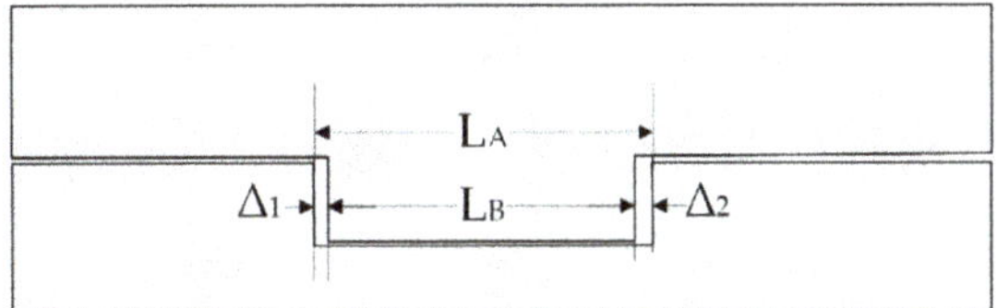

Figure 1. Coordination accuracy of a keyway assembly.

The coordination accuracy is

$$\nabla_{AB} =_1 +_2 = L_A - L_B,\tag{1}$$

It can be seen that the coordination accuracy is the amount of the allowance on a certain dimension. In this way, the assembly coordination of a single dimension is well presented by coordination accuracy such as angle, length, etc. However, it is not suitable for complicated assembly. Therefore, the concept should be extended to pose allowance space and the space can be predicted by digital measurement data during large-scale assembly. This space is named the assembly coordination space, which is the ability of pose variation under the condition of assembly accuracy requirements, as Figure 2 shows.

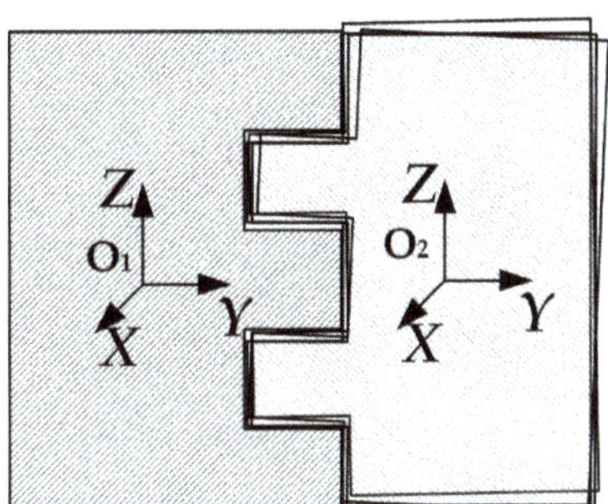

Figure 2. The pose variations under the condition of assembly accuracy requirements.

The parts of the assembly are divided into the reference part and the align part. The reference part is the fixed part during assembly and the align part will move to the target pose by the pose adjustment tooling. Assume that the primary measurement data of the two parts are

$$\begin{cases} P^R = \begin{bmatrix} p_1^R & p_2^R & \cdots & p_n^R \end{bmatrix} \\ P^A = \begin{bmatrix} p_1^A & p_2^A & \cdots & p_n^A \end{bmatrix} \end{cases},\tag{2}$$

where P^R and P^A are the point sets of the reference part and align part, separately, where p_1^R, etc. and p_1^A, etc. are the points of the sets P^R and P^A, respectively. The two parts are separated first. According to the least-squares method, the optimal assembly pose can be calculated by

$$\begin{cases} P^R \approx RP^A + T \\ e = \min\left\{ \sum_{i=1}^{n} \| Rp_1^A + T - P^R \|^2 \right\} \end{cases},\tag{3}$$

where R is the rotation matrix, T is the movement matrix, and e is the minimum residual sum of squares. The singular value decomposition method [24] is taken to calculate the parameter R and T. Then, the optimal pose of the align part based on the least-squares method is

$$\omega_0 = \begin{bmatrix} R & T \\ 0 & 1 \end{bmatrix},$$ (4)

The assembly deviation can be predicted by the pose ω_0 of the align part. The key assembly characteristics (KAC) [25] are the important geometric structures that have key influences on assembly quality. They are described by measurement data and some dimensions that are not necessary to be measured.

$$K = \{P, G\},$$ (5)

where K is the parameters of a KAC, P is the measurement data, G is the dimensions that are not necessary to be measured. The KACs have an irregular distribution in space during large-scale assembly. As shown in Figure 3, the wing-fuselage assembly is completed by 4 pairs of joints. There are four assembly accuracy requirements on each pair of joints: Two on coaxialities and two on clearances.

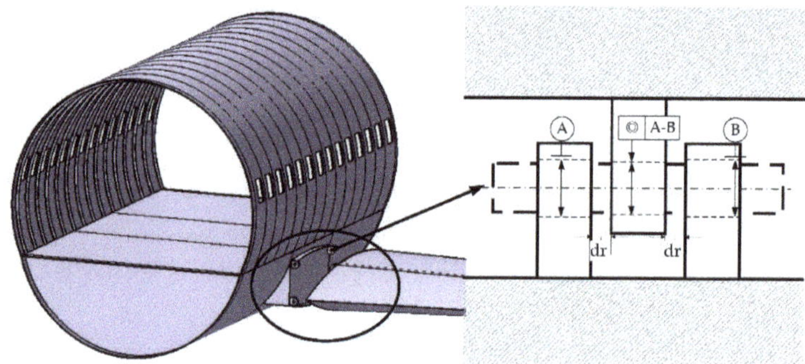

Figure 3. Wing-fuselage assembly.

The KACs are restrained by assembly accuracy requirements. The assembly accuracy is described as

$$T_j^i = f_j^i\left(K_i^R, K_i^A\right),$$ (6)

where K_i^R is the parameters of the ith KAC on the reference part (fuselage), K_i^A is the parameters of the ith KAC on the align part (wing), T_j^i is the jth assembly accuracy of the ith KAC, and f_j^i is the mapping from parameters to the assembly accuracy. The assembly accuracy should meet the requirements of assembly accuracy, which is formulated in Equation (7)

$$T_j^i \in \left[T_j^{i-min}, T_j^{i-max}\right],$$ (7)

where T_j^{i-min} and T_j^{i-max} are the ranges of T_j^i. Substitute Equation (6) into Equation (7):

$$f_j^i\left(K_i^R, K_i^A\right) \in \left[T_j^{i-min}, T_j^{i-max}\right],$$ (8)

For the m assembly accuracy requirements on n KACs of the assembly, the constraint equations can be expressed as

$$\forall f_j^i\left(K_i^R, K_i^A\right) \in \left[T_j^{i-min}, T_j^{i-max}\right], i \in [1, n], j \in [1, j_i],$$ (9)

where j_i is the assembly accuracy requirement number of the ith KAC, $m = \sum_{i=1}^{n} j_i$. When all KACs satisfy their assembly accuracy requirements, the pose is a valid pose to be aligned.

As shown in Figure 4, the valid pose may not be the only one that satisfies all assembly accuracy requirements. Therefore, the adjacent poses of the primary pose shown in Figure 4a can be analyzed. A small displacement torsor (SDT) [26] represents a tiny rigid body's pose variation. It is described as

$$\omega_\Delta = (x_\Delta, y_\Delta, z_\Delta, \alpha_\Delta, \beta_\Delta, \gamma_\Delta), \tag{10}$$

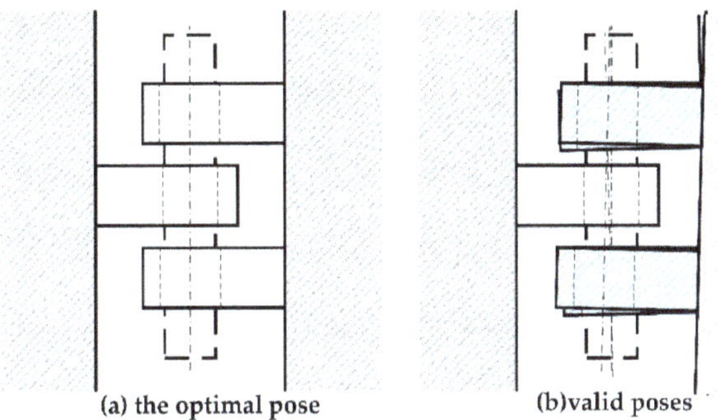

(a) the optimal pose   (b)valid poses

Figure 4. Valid poses of wing-fuselage assembly.

The homogeneous transformation matrix of an SDT is

$$\omega_\Delta^H = \begin{bmatrix} C\gamma_\Delta C\beta_\Delta & -S\gamma_\Delta C\beta_\Delta & S\beta_\Delta & x_\Delta \\ S\gamma_\Delta C\alpha_\Delta + C\gamma_\Delta S\beta_\Delta S\alpha_\Delta & C\gamma_\Delta C\alpha_\Delta - S\gamma_\Delta S\beta_\Delta S\alpha_\Delta & -C\beta_\Delta S\alpha_\Delta & y_\Delta \\ S\gamma_\Delta S\alpha_\Delta - C\gamma_\Delta S\beta_\Delta C\alpha_\Delta & C\gamma_\Delta S\alpha_\Delta + S\gamma_\Delta S\beta_\Delta C\alpha_\Delta & C\beta_\Delta C\alpha_\Delta & z_\Delta \\ 0 & 0 & 0 & 1 \end{bmatrix} \approx \begin{bmatrix} 1 & -\gamma_\Delta & \beta_\Delta & x_\Delta \\ \gamma_\Delta & 1 & -\alpha_\Delta & y_\Delta \\ -\beta_\Delta & \alpha_\Delta & 1 & z_\Delta \\ 0 & 0 & 0 & 1 \end{bmatrix}, \tag{11}$$

where S is sin, C is cos, $\lim_{\alpha_\Delta \to 0} C\alpha_\Delta = 1$, $\lim_{\alpha_\Delta \to 0} S\alpha_\Delta = \alpha_\Delta$, and $\lim_{\alpha_\Delta, \beta_\Delta \to 0} S\alpha_\Delta S\beta_\Delta = 0$. The change in point $p = (x, y, z)$ after a slight change in the rigid body's pose is

$$p'_{ex} = [x \, y \, z \, 1] \begin{bmatrix} 1 & -\gamma_\Delta & \beta_\Delta & x_\Delta \\ \gamma_\Delta & 1 & -\alpha_\Delta & y_\Delta \\ -\beta_\Delta & \alpha_\Delta & 1 & z_\Delta \\ 0 & 0 & 0 & 1 \end{bmatrix} = \begin{bmatrix} x + x_\Delta + \beta_\Delta \cdot z - \gamma_\Delta \cdot z \\ y + y_\Delta - \alpha_\Delta \cdot z + \gamma_\Delta \cdot x \\ z + z_\Delta + \alpha_\Delta \cdot y - \beta_\Delta \cdot x \\ 1 \end{bmatrix}^T, \tag{12}$$

Then, the assembly accuracy would be

$$T_j^{i-\omega_\Delta} = f_j^i\left(K_i^R, K_i^A \omega_\Delta^H\right) = f_j^i\left(P_i^R, G_i^R, P_i^A \omega_\Delta^H, G_i^A\right), \tag{13}$$

On this pose, if the assembly accuracy requirements are still satisfied as

$$\forall f_j^i\left(K_i^R, K_i^A \omega_\Delta^H\right) \in \left[T_j^{i-min}, T_j^{i-max}\right], i \in [1, n], j \in [1, j_i], \tag{14}$$

the pose is still a valid pose. The coordination space model can, hence, be expressed as

$$\varnothing_{CS} = \left\{\omega_\Delta \middle| \forall f_j^i\left(K_i^R, K_i^A \omega_\Delta^H\right) \in \left[T_j^{i-min}, T_j^{i-max}\right], i \in [1, n], j \in [1, j_i]\right\}, \tag{15}$$

where $\varnothing_{CS}$ is the coordination space, which is the whole pose variation space under the condition of assembly accuracy requirements.

3. Assemblability Analysis Based on Coordination Space Model

Assemblability refers to the geometric consistency of the matching geometric structures of the two assembling parts. It can be judged whether the assembly can directly be carried out by assemblability prediction.

The assemblability is good if the coordination space is greater than 0, which means at least one pose conforms to Equation (15). Otherwise, the assemblability is bad. Therefore, the assemblability analysis flow is shown in Figure 5.

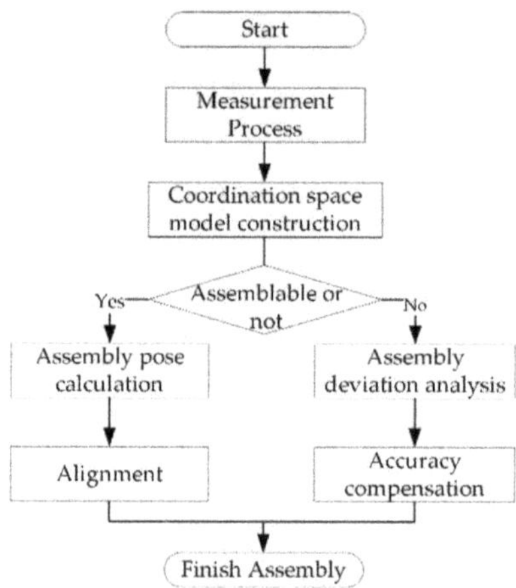

Figure 5. Assemblability analysis process.

Firstly, the KACs are measured by a laser tracker or other digital measurement devices. Then, the coordination space model is constructed based on the assembly accuracy requirements. The volume of the coordination space is solved to judge whether it is assemblable. It will be assemblable when $\varnothing_{CS}$ is greater than 0. The assembly can be executed by calculating the optimal pose and aligning the parts. It will be uncoordinated when $\varnothing_{CS}$ is 0. Then, the assembly deviation should be analyzed and compensated to make it assemblable.

The solution process of the coordination space is based on the Monte Carlo method:

1. Calculate the optimal pose based on the least-squares method.
2. According to the dimensions and assembly accuracy requirements, a maximum pose space is assumed, as Equation (16) shows. All poses out of the space are not valid for any assembly accuracy requirements.

$$\omega_{\mathrm{d}} : (-x_{\mathrm{d}}, -y_{\mathrm{d}}, -z_{\mathrm{d}}, -\alpha_{\mathrm{d}}, -\beta_{\mathrm{d}}, -\gamma_{\mathrm{d}}) \rightarrow (x_{\mathrm{d}}, y_{\mathrm{d}}, z_{\mathrm{d}}, \alpha_{\mathrm{d}}, \beta_{\mathrm{d}}, \gamma_{\mathrm{d}}), \tag{16}$$

3. Generate a random SDT uniformly for n_T times and check the SDTs by Equation (15).
4. If n_j of n_T SDTs are valid, the coordination space is

$$\varnothing_{CS} = \frac{n_j}{n_T} 64 x_{\mathrm{d}} y_{\mathrm{d}} z_{\mathrm{d}} \alpha_{\mathrm{d}} \beta_{\mathrm{d}} \gamma_{\mathrm{d}}, \tag{17}$$

In the case of incoordination, the coordination space should be further analyzed. According to Equation (15), the coordination space is the intersection of KAC's constraint equations. All constraints are divided by KACs. Equation (15) will be translated to

$$\left\{ \omega_\Delta \left| \begin{array}{l} \forall f_j^1\left(K_1^R, K_1^A \omega_\Delta^H\right) \in \left[T_j^{1-min}, T_j^{1-max}\right], j \in [1, j_1], \\ \forall f_j^2\left(K_2^R, K_2^A \omega_\Delta^H\right) \in \left[T_j^{2-min}, T_j^{2-max}\right], j \in [1, j_2], \\ \cdots \\ \forall f_j^n\left(K_n^R, K_n^A \omega_\Delta^H\right) \in \left[T_j^{n-min}, T_j^{n-max}\right], j \in [1, j_n] \end{array} \right. \right\}, \tag{18}$$

Let

$$\varnothing_i^{KAC} = \left\{ \omega_\Delta^{i-KAC} \left| \forall f_j^i\left(K_i^R, K_i^A \omega_\Delta^H\right) \in \left[T_j^{i-min}, T_j^{i-max}\right], j \in [1, j_i] \right. \right\}, \tag{19}$$

where $\varnothing_i^{KAC}$ is the *KAC* coordination space formed by the assembly accuracy requirements of a *KAC*, and ω_Δ^{i-KAC} is an SDT in the *KAC* coordination space. The $\varnothing_{CS}$ would be

$$\varnothing_{CS} = \bigcap_{i=1}^n \varnothing_i^{KAC}, \tag{20}$$

The relationship between the KAC coordination space and the assembly coordination space is shown in Figure 6a.

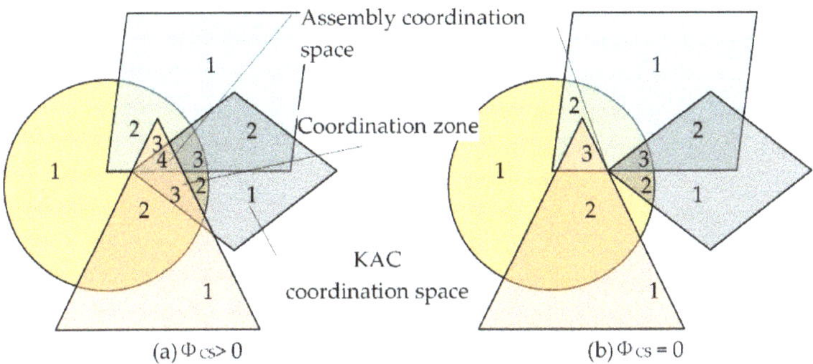

Figure 6. The relationship between the part and feature PFS.

Figure 6b shows the status of the *KAC* coordination space when the assembly is uncoordinated. Each area of the same color represents a *KAC* coordination space. The divided zone is named the coordination zone. The accuracy compensation method is needed to improve the assemblability.

According to Figure 6b, set the union of *KAC* coordination space as a union coordination space. It is formulated as

$$\varnothing_{UCS} = \bigcup_{i=1}^n \varnothing_i^{KAC}, \tag{21}$$

where $\varnothing_{UCS}$ is the union coordination space. In the union coordination space, all poses are valid for some KACs but not valid for all. Some divided zones are valid for more KACs than others, e.g., the two zones marked with 3 are better than those marked with 1 or 2. The marked number is named the coordination zone index, which is the valid KACs' number in the coordination zone. If a pose in the zone marked with 3 is selected, only one KAC needs to be compensated. In this way, an assemblability optimization method is put forward by selecting a coordination zone with larger volume and KAC number. The larger volume means a better geometric consistency, and the larger KAC number means fewer KACs need to be compensated.

4. Assemblability Optimization Based on the Union Coordination Space

The accuracy compensation process is time- and effort- consuming [22] when the assemblability is poor. For example, it needs programming, clamping, tool setting, machining, loosen clamping, and other steps when finishing a KAC with cutting. Therefore, reducing the number of KACs to be processed is an effective means to improve the assembly efficiency in many cases. The optimal pose is usually obtained under the condition of optimal assembly accuracy. If each unqualified KAC is compensated one by one under the optimal pose, more work may be needed and the assembly quality might not be good, due to the unknown assembly quality after accuracy compensation. If the assembly quality is bad after compensation, there are no alternative compensation schemes based on the least-squares method. Therefore, the assemblability optimization method is proposed to solve the incoordination problem. The key to optimize the assemblability is whether there is one or more coordination zones that can satisfy assembly accuracy requirements with fewer KACs to be compensated and a better or approximate volume of coordination space.

The coordination zone index shows the valid KACs in the certain coordination zone. The total number of all KACs is n_{KAC}. The incoordination zone index shows the number of uncoordinated KACs in the coordination zone. Their relationship is

$$n_{IZI} = n_{KAC} - n_{CZI}, \tag{22}$$

where n_{IZI} is the incoordination zone index and n_{CZI} is the coordination zone index. If the accuracy of uncoordinated KACs is compensated well in the coordination zone, this coordination zone will change to the assembly coordination space, as Figure 7 shows.

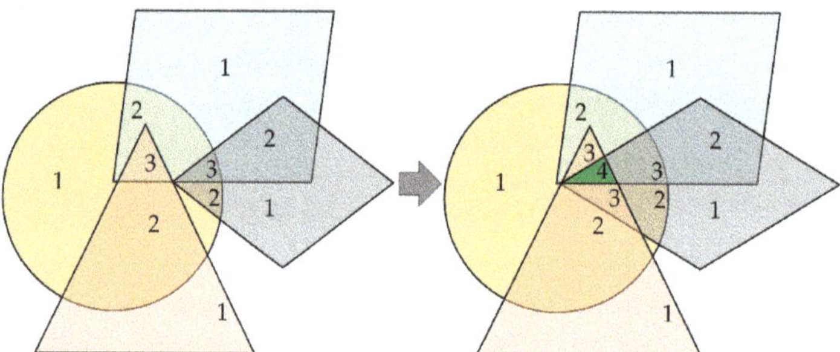

Figure 7. The coordination zone variation process by accuracy compensation.

In this way, each coordination zone can be analyzed to check whether it is good to be compensated or not. Two indicators of the coordination zone should be analyzed, one is the incoordination zone index, and the other is the volume of the coordination zone. The Monte Carlo method of Section 3 is improved to judge the state of each coordination zone one by one, and the optimal assemblability optimization schemes of the coordination zone are selected for recording.

The solution process based on the Monte Carlo method is as follows:

1. Solve the optimal pose of the align part;
2. Set a pose space as the pose boundary as shown by the square box of Figure 8;
3. Generate a random SDT in the pose space;
4. According to Equation (19), judge which KAC equations are satisfied (coordination zone index) and which are not (incoordination zone index);
5. Cluster the analysis results of each SDT. The SDTs in the same coordination zone are clustered together;
6. Put the clustered results into the data structure of Equation (23). The KAC number to be compensated is the incoordination zone index. Select the scheme with a better KAC number and space volume of the coordination zone.

$$\left\{ \Gamma | \Gamma_i = \left(n_{IZI}, V_{CZ}, b_f, s_\omega \right), i < \Gamma_{num} \right\}, \tag{23}$$

where Γ_i is the *i*th scheme, n_{IZI} is the incoordination zone index, V_{CZ} is the space amount of the coordination zone, b_f is the information of uncoordinated KACs, s_ω is the SDT set, and Γ_{num} is the max number of the schemes.

7. Calculate the center SDT of the SDTs in the selected scheme. The assembly deviation of target features under the SDT is analyzed and the accuracy compensation is carried out.

$$\omega_\Delta^c = \frac{1}{n_s} \sum_{i=1}^{n_s} \omega_{\Delta i}, \tag{24}$$

where ω_Δ^c is the center SDT, n_s is the SDT number of s_ω, and $\omega_{\Delta i}$ is an SDT of s_ω. All assembly accuracies on ω_Δ^c are calculated. Then, the deviations on the excessive KACs will be compensated.

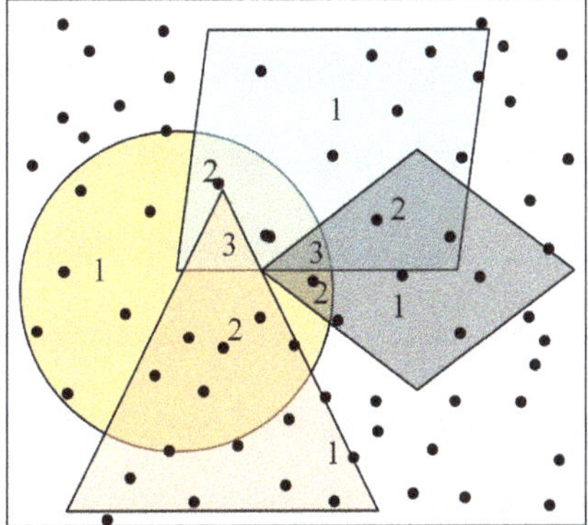

Figure 8. Random sampling by Monte Carlo method.

According to Equations (13) and (24), the compensation amount would be

$$C_j^i = T_j^{i-\omega_\Delta^c} - T_j^{i-opt} = f_j^i\left(K_i^R, K_i^A \omega_\Delta^c\right) - T_j^{i-opt}, \tag{25}$$

where C_j^i is the compensation amount of the *j*th assembly accuracy requirement of the *i*th *KAC*, $T_j^{i-\omega_\Delta^c}$ is the assembly accuracy on the SDT ω_Δ^c, and T_j^{i-opt} is the optimal value of the assembly accuracy.

5. Case Study

5.1. Space Manipulator Assembly

The space manipulator is fixed on the spacecraft, which needs a high assembly accuracy to guarantee the stability when the spacecraft is flying. The assembly is executed by shaft and hole connectors, which are shown in Figure 9a. The connector is shown in Figure 9b. The manipulator is the align part and the spacecraft is the reference part.

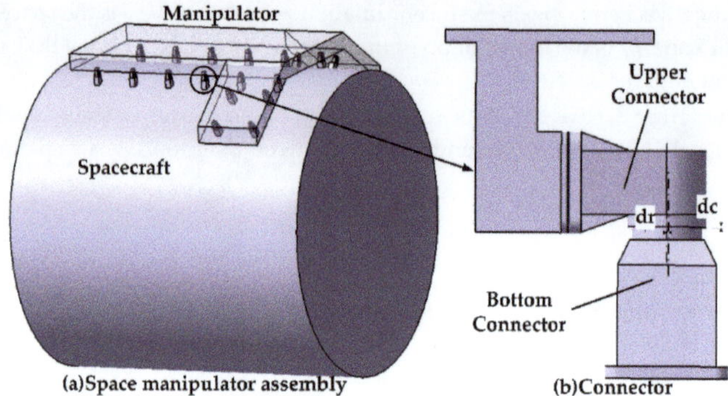

Figure 9. Space manipulator assembly.

The upper connector is fixed on the manipulator, and the bottom connector is fixed on the spacecraft. The KACs are the assembly of the connectors. Due to the slight deformation of the spacecraft and the installation error of the bottom connectors, it is difficult for the connectors to accurately assemble at one time during the assembly of the spacecraft and the manipulator. In the original assembly process, it is necessary to try the assembly first, measure the assembly deviation of the clearance and coaxiality of each pair of connectors, make the accuracy compensation, and retry the assembly to ensure the assembly quality. The assembly takes a long time and the connectors are not convenient to be operated on the spacecraft. Therefore, the laser tracker is used to measure the connectors between the spacecraft and the manipulator. The methods in Sections 2 and 3 are taken to evaluate the assemblability based on the measurement data. The method in Section 4 is used to find the key connectors to make the accuracy compensation. The assembly is carried out after the accuracy compensation. In this way, the assembly quality is better guaranteed and the assembly efficiency is improved. The flow of the proposed method and the comparison with the original method are shown in Figure 10.

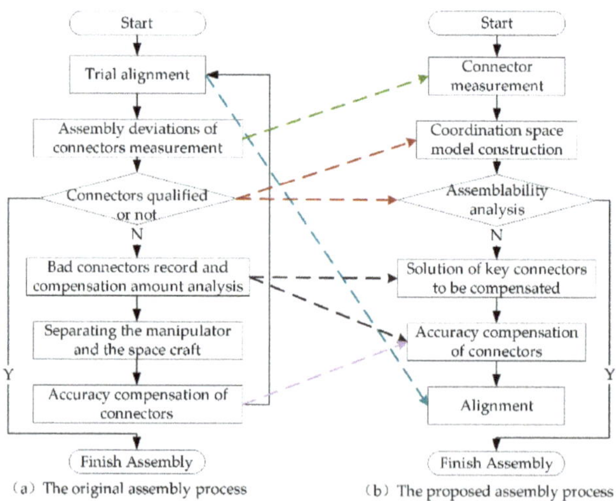

Figure 10. The comparison of the original and proposed assembly processes.

As shown in Figure 10, the assembly process is developed toward digital measurement and analysis. The results obtained in the actual assembly and inspection are replaced by the analysis of the measurement data. Therefore, some unnecessary assembly processes are eliminated and the possibility of repeated trial assembly is greatly reduced.

The assembly accuracy requirements of the connector are coaxiality dr and clearance dc on the matching surface, as shown in Figure 9b. The coaxiality requirement is 0.2 mm, and the clearance requirement is 0.1 mm. Assembly accuracy is compensated by gasket compensation, finishing, or position movement according to the deviation.

5.2. Coordination Space Model

The measurement of the connector is based on the measurement auxiliary tool, which is shown in Figure 11.

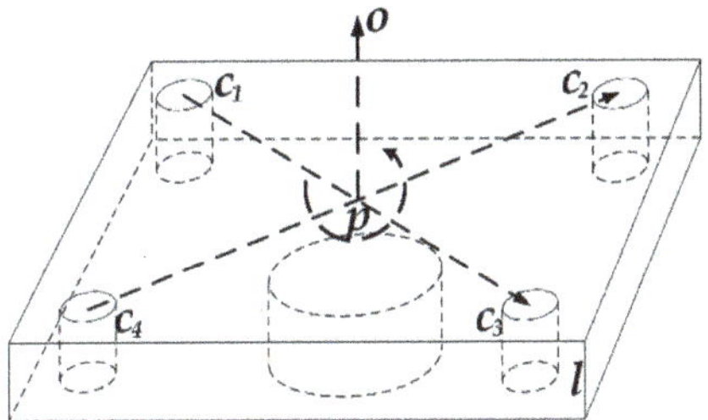

Figure 11. Measurement auxiliary tool.

After inserting the shaft into the corresponding hole, measure the four holes of the measurement auxiliary tool. The measurement data are processed as the position p and orientation $\vec{po}$.

$$\begin{cases} \vec{po} = \frac{(c_3-c_1)\times(c_4-c_2)}{|(c_3-c_1)\times(c_4-c_2)|} \\ p = \frac{1}{4}\sum c_i - l\cdot\vec{po} \end{cases}, \tag{26}$$

where c_1, c_2, c_3, and c_4 are the points measured by the laser tracker.

As shown in Figure 12, the clearance dc and the coaxiality dr of a connector are

$$\begin{cases} dr = \left|\vec{p_2p_1}\cdot sin\theta_2\right| \\ dc = \left|\vec{p_1p_2}\right|\cdot cos\theta_2 + r\cdot sin(\theta_2 - \theta_1) \end{cases}, \tag{27}$$

where θ_1 and θ_2 are the angles between $\vec{p_1p_2}$ and $\vec{p_1o_1}$ or $\vec{p_1o_2}$; θ_1 can be calculated by $\theta_1 = arccos\left(\vec{p_1p_2}\cdot\vec{p_1o_1}/\left|\vec{p_1p_2}\right|\left|\vec{p_1o_1}\right|\right)$ and, similarly, θ_2 can be calculated by the same way; and r is the radius of the matching surface, which is 15 mm. There are 20 connectors to be guaranteed at the same time.

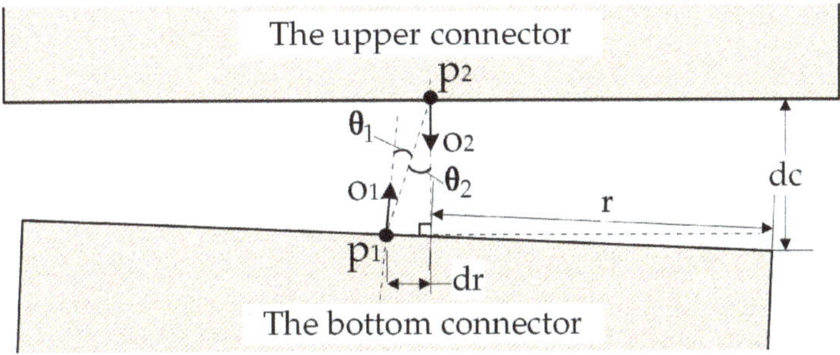

Figure 12. Assembly geometric constraints analysis.

Therefore, the coordination space model is

$$\left\{ \omega_\Delta \left| \begin{array}{c} \left| \vec{p_2}^{i-\omega_\Delta} p_1^i \omega_\Delta^H \cdot \sin\theta_2^{i-\omega_\Delta} \right| < 0.2, \\ 0 < \left| p_1^i \vec{p_2}^{i-\omega_\Delta} \right| \cdot \cos\theta_2^{i-\omega_\Delta} + r \cdot \sin\left(\theta_2^{i-\omega_\Delta} - \theta_1^{i-\omega_\Delta}\right) < 0.1, \\ i \in [1, 20] \end{array} \right. \right\}, \tag{28}$$

where ω_Δ is the random SDT based on the optimal pose derived from the least-squares method, and $p_2^{i-\omega_\Delta}$ is the parameters of Equation (27) changed by ω_Δ according to Equation (12), which are listed in Equation (29):

$$\left\{ \begin{array}{l} \vec{p_2}^{i-\omega_\Delta} = p_2^i \omega_\Delta^H \\ \vec{p_2}^{i-\omega_\Delta} o_2^{i-\omega_\Delta} = \dfrac{(c_3^i - c_1^i) \times (c_4^i - c_2^i)}{\left| (c_3^i - c_1^i) \times (c_4^i - c_2^i) \right|} \\ \theta_1^{i-\omega_\Delta} = arccos\left(\vec{p_1^i p_2^{l-\omega_\Delta}} \cdot \vec{p_1^i o_1^l} \Big/ \left| \vec{p_1^i p_2^{l-\omega_\Delta}} \right| \left| \vec{p_1^i o_1^l} \right| \right) \\ \theta_2^{i-\omega_\Delta} = arccos\left(\vec{p_1^i p_2^{l-\omega_\Delta}} \cdot \vec{p_2^{l-\omega_\Delta} o_2^{l-\omega_\Delta}} \Big/ \left| \vec{p_1^i p_2^{l-\omega_\Delta}} \right| \left| \vec{p_2^{l-\omega_\Delta} o_2^{l-\omega_\Delta}} \right| \right) \end{array} \right., \tag{29}$$

5.3. Assemblability Analysis

Part of the raw data is listed in Table 1. All the measurement data are listed in Appendix A, Table A1.

Table 1. Part of the raw measurement data.

Spacecraft			Manipulator		
x/mm	y/mm	z/mm	x/mm	y/mm	z/mm
26.060	26.037	−192.674	222.509	2274.256	403.730
25.941	−25.881	−192.645	222.471	2222.338	403.647
−25.972	−25.945	−192.558	170.389	2222.349	403.666
−25.973	26.032	−192.581	170.520	2274.362	403.722

The least-squares method is taken to calculate the optimal pose and the deviations on the optimal pose. The deviations of the connectors are listed in Table 2 calculated by Equation (27).

Table 2. Assembly deviation prediction by least-squares method.

No.	dr/mm	dc/mm	No.	dr/mm	dc/mm
1	0.154	0.043	11	0.171	−0.012
2	0.150	−0.027	12	0.338	0.002
3	0.205	0.072	13	0.286	−0.010
4	0.196	0.042	14	0.530	−0.274
5	0.191	0.012	15	0.247	0.033
6	0.209	0.083	16	0.165	−0.029
7	0.066	0.045	17	0.188	0.033
8	0.195	−0.025	18	0.402	−0.154
9	0.118	−0.006	19	0.086	0.183
10	0.298	−0.010	20	0.386	−0.028

It can be seen that 10 connectors need to be adjusted or repaired based on the least-squares method. The coordination space is 0 at the optimal pose based on the method in Section 3, which means it cannot be assembled directly. Therefore, the assemblability should be optimized.

5.4. Assemblability Optimization

The proposed method in Section 4 is taken to find the accuracy compensation schemes. The results are shown in Figure 13.

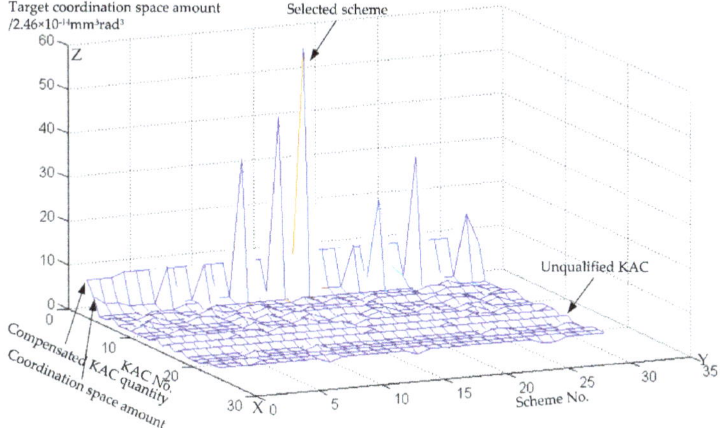

Figure 13. The accuracy compensation schemes.

Figure 13 shows the accuracy compensation schemes. The first point on the X axis is the KAC quantity to be compensated. The second point is the volume of the coordination zone of the scheme. The latter ones are the number of KACs. The Y axis is the scheme number. The Z axis is the value of the X axis. Seven connectors need to be adjusted to complete assembly in scheme 1. Finally, scheme 18, which needs nine connectors to be compensated, is taken by considering the assembly quality. The coordination space of the scheme is 56d. d is the volume of the maximum pose space divided by random times. In this case, d is 2.46×10^{-14} mm³rad³. The KAC number to be compensated is 5, 6, 8, 10, 12, 14, 18, 19, and 20. The center SDT of the coordination zone in scheme 18 is $\left(-0.0363 \text{ mm}, 0.0210 \text{ mm}, 0.0098 \text{ mm}, 2.52 \times 10^{-6} \text{ rad}, 1.64 \times 10^{-6} \text{ rad}, -2.35 \times 10^{-6} \text{ rad}\right)$.

The deviation is calculated under the center SDT listed in Table 3.

Table 3. The deviations of the connectors.

No.	dr/mm	dc/mm	No.	dr/mm	dc/mm
1	0.145	0.034	11	0.150	−0.036
2	0.188	−0.031	12	0.351	−0.013
3	0.169	0.064	13	0.142	−0.022
4	0.151	0.031	14	0.391	−0.286
5	0.230	0.004	15	0.133	0.016
6	0.249	0.070	16	0.031	−0.039
7	0.025	0.037	17	0.056	0.032
8	0.222	−0.034	18	0.204	−0.145
9	0.162	−0.021	19	0.204	0.179
10	0.255	−0.029	20	0.233	−0.027

After simulation accuracy compensation for the above nine connectors, which is in bold and italics in Table 3 (the proposed method), the coordination space is 101d, which is greater than 0. The average coaxiality is 0.068 mm and the average clearance is 0.014 mm. The assemblability is good and the assembly can be executed directly.

After simulation accuracy compensation for the above 10 connectors, which is in bold and italics in Table 2 (the least square method), the coordination space is 9d. The average coaxiality is 0.084 mm and the average clearance is 0.014 mm. The assemblability is good but the assembly quality on coaxiality is worse.

The result shows that the proposed method will generate a better accuracy compensation scheme with less workload and better assembly quality, which improves the assemblability.

The measurement and connector adjustment process took about 8 h during the assembly. The pose adjustment process took about 2 h. Therefore, it took about 10 h in total based on the proposed method. The original assembly process took more than 20 h because the first trial assembly and accuracy compensation process cannot realize the re-trial assembly smoothly. Three or four times the assembly are needed to guarantee the assembly quality.

6. Discussion

Compared to the previous research, the major contributions in this paper are listed as follows: (1) The concept of assemblability and coordination accuracy in the design/drawing stage are extended into the measurement-assisted assembly. (2) An assemblability analysis method based on the measurement data and the coordination space model is proposed for predicting the key assembly deviations. (3) The accuracy compensation methods based on the optimal pose might lead to more workload and worse assemblability. Therefore, an assemblability optimization method is proposed for less workload and better assembly quality. In addition, the space manipulator assembly is taken as an example. The result shows that the proposed method can optimize the assemblability with less workload and better assembly quality compared to the accuracy compensation method based on the optimal pose.

The assemblability optimization method based on accuracy compensation improves the ability to detect assembly problems in advance, which will benefit the automation assembly. Further, the coordination space model and the small displacement torsor are useful for analyzing the assemblability and optimizing the tolerances in the design/drawings phase, but the assemblability optimization method is not useful. In the implementation of the method, high-precision digital measurement equipment are needed. Measurement uncertainty will affect the reliability of the final results.

Future research include evaluating the influence of the measurement uncertainty on the coordination space model. Then, the uncertainty of pose adjustment should be taken into consideration compared to the volume of the coordination space to judge the feasibility of automatic pose adjustment.

Author Contributions: Conceptualization, Z.C.; methodology, Z.C.; software, Z.C.; validation, Z.C.; formal analysis, Z.C.; investigation, Z.C.; resources, Z.C.; data curation, Z.C.; writing—original draft preparation, Z.C.; writing—review and editing, F.D.; visualization, Z.C.; supervision, F.D.; project administration, F.D.; funding acquisition, F.D. All authors have read and agreed to the published version of the manuscript.

Funding: This research was funded by the Basic Scientific Research, grant number JCKY2016206B009 and the Marine Power Research & Development (MPRD).

Conflicts of Interest: The authors declare no conflict of interest.

Appendix A

Table A1. Raw measurement data.

Spacecraft			Spacecraft			Manipulator			Manipulator		
x/mm	y/mm	z/mm	x/mm	y/mm	z/mm	x/mm	y/mm	z/mm	x/mm	y/mm	z/mm
26.060	26.037	−192.674	2525.979	25.966	−192.595	222.509	2274.256	403.730	2722.472	2274.418	403.607
25.941	−25.881	−192.645	2525.879	−26.066	−192.588	222.471	2222.338	403.647	2722.406	2222.370	403.680
−25.972	−25.945	−192.558	2473.905	−25.939	−192.634	170.389	2222.349	403.666	2670.395	2222.418	403.611
−25.973	26.032	−192.581	2473.927	26.032	−192.648	170.520	2274.362	403.722	2670.399	2274.318	403.725
25.964	−1103.891	−192.620	2525.886	−1103.863	−192.642	222.163	1144.197	403.734	2722.221	1144.019	403.649
25.995	−1155.853	−192.533	2525.963	−1155.983	−192.687	222.172	1092.171	403.594	2722.191	1092.064	403.681
−26.108	−1155.891	−192.633	2473.910	−1155.938	−192.617	170.261	1092.242	403.670	2670.236	1092.093	403.632
−26.138	−1103.914	−192.600	2473.997	−1103.946	−192.532	170.240	1144.209	403.636	2670.134	1144.022	403.682
526.046	26.182	−192.691	25.984	371.512	−292.868	722.540	2274.127	403.704	222.310	2619.620	303.466
526.032	−25.859	−192.578	26.082	328.937	−322.879	722.525	2222.138	403.647	222.406	2577.169	273.462
474.044	−25.874	−192.632	−25.948	328.968	−322.927	670.477	2222.107	403.660	170.271	2577.228	273.409
473.999	26.048	−192.660	−25.989	371.392	−292.860	670.519	2274.160	403.749	170.259	2619.691	303.502
525.926	−1103.994	−192.665	26.306	779.693	−581.183	722.553	1144.152	403.622	222.373	3028.126	15.112
525.941	−1156.028	−192.670	26.328	737.195	−611.193	722.582	1092.046	403.733	222.475	2985.706	−14.926
473.906	−1155.956	−192.534	−25.677	737.131	−611.127	670.483	1092.007	403.687	170.440	2985.796	−14.932
474.001	−1103.938	−192.549	−25.718	779.531	−581.201	670.468	1144.107	403.701	170.438	3028.159	15.042
1025.900	25.892	−192.696	526.057	371.467	−292.910	1222.170	2274.325	403.687	722.232	2619.736	303.446
1025.991	−26.042	−192.675	525.980	329.046	−322.917	1222.300	2222.253	403.652	722.264	2577.274	273.375
973.956	−26.131	−192.596	473.953	329.017	−322.836	1170.188	2222.222	403.641	670.255	2577.165	273.441
973.988	25.990	−192.571	473.989	371.499	−292.943	1170.178	2274.330	403.697	670.274	2619.662	303.464
1026.139	−1104.097	−192.694	526.004	779.757	−581.219	1222.329	1144.177	403.676	722.338	3027.977	14.999
1026.085	−1156.050	−192.740	525.882	737.293	−611.181	1222.380	1092.031	403.649	722.348	2985.465	−14.896
974.124	−1156.044	−192.605	474.011	737.365	−611.149	1170.291	1092.073	403.617	670.334	2985.461	−14.882
974.105	−1104.011	−192.634	474.022	779.818	−581.198	1170.331	1144.076	403.725	670.270	3028.053	15.054
1525.925	26.129	−192.717	25.959	−1458.921	−322.909	1722.341	2274.278	403.722	222.442	789.300	273.358
1525.908	−25.848	−192.628	26.022	−1501.395	−292.846	1722.343	2222.377	403.704	222.364	746.770	303.402
1473.914	−25.921	−192.601	−26.061	−1501.356	−292.887	1670.349	2222.277	403.637	170.403	746.817	303.352
1473.942	26.171	−192.663	−25.980	−1458.862	−322.971	1670.315	2274.352	403.667	170.462	789.233	273.423
1526.063	−1104.006	−192.639	26.112	−1867.403	−611.275	1722.264	1144.106	403.701	222.487	380.493	−14.929
1525.951	−1155.886	−192.590	25.999	−1909.862	−581.157	1722.267	1092.232	403.641	222.479	338.017	15.121

Table A1. *Cont.*

	Spacecraft			Spacecraft			Manipulator			Manipulator	
x/mm	y/mm	z/mm	x/mm	y/mm	z/mm	x/mm	y/mm	z/mm	x/mm	y/mm	z/mm
1473.970	−1155.991	−192.610	−25.945	−1909.883	−581.133	1670.247	1092.136	403.697	170.454	337.946	15.019
1474.029	−1103.917	−192.536	−25.952	−1867.415	−611.214	1670.338	1144.104	403.578	170.502	380.501	−15.022
2026.006	25.807	−192.610	525.901	−1459.048	−322.918	2222.301	2274.138	403.623	722.541	789.350	273.516
2026.008	−26.174	−192.602	525.980	−1501.511	−292.844	2222.194	2222.153	403.685	722.525	746.885	303.481
1974.072	−26.202	−192.597	474.025	−1501.571	−292.828	2170.273	2222.082	403.688	670.449	746.930	303.374
1973.940	25.758	−192.615	473.983	−1459.106	−322.849	2170.301	2274.070	403.707	670.465	789.270	273.409
2025.839	−1103.981	−192.592	525.886	−1867.434	−611.169	2222.555	1144.297	403.746	722.630	380.599	−14.830
2025.808	−1155.974	−192.526	525.806	−1909.943	−581.114	2222.521	1092.226	403.623	722.554	338.263	15.129
1973.868	−1155.897	−192.571	473.872	−1909.917	−581.157	2170.597	1092.349	403.654	670.639	338.232	15.073
1973.894	−1103.986	−192.651	473.840	−1867.343	−611.211	2170.608	1144.349	403.683	670.507	380.619	−14.907

References

1. Zhou, F.; Xue, H.; Zhou, W.; Xu, G. Key technology and its improvement of aircraft digital flexible assembly. *Aeronaut. Manuf. Technol.* **2006**, *9*, 30–35. (In Chinese)
2. Maropoulos, P.G.; Muelaner, J.E.; Summers, M.D.; Martin, O.C. A new paradigm in large-scale assembly—Research priorities in measurement assisted assembly. *Int. J. Adv. Manuf. Technol.* **2014**, *70*, 621–633. [CrossRef]
3. Marguet, B.; Ribere, B. Measurement-assisted assembly applications on airbus final assembly lines. *SAE Tech. Pap.* **2003**, *1*, 2950.
4. Chen, Z.; Du, F.; Tang, X. Position and orientation best-fitting based on deterministic theory during large scale assembly. *J. Intell. Manuf.* **2015**, *29*, 827–837. [CrossRef]
5. Li, S.; Deng, Z.; Zeng, Q.; Huang, X. A coaxial alignment method for large aircraft component assembly using distributed monocular vision. *Assem. Autom.* **2018**, *38*, 437–449. [CrossRef]
6. Wang, Q.; Dou, Y.; Li, J.; Ke, Y.; Keogh, P.; Maropoulos, P.G. An assembly gap control method based on posture alignment of wing panels in aircraft assembly. *Assem. Autom.* **2017**, *37*, 422–433. [CrossRef]
7. Sukhan, L.; Chunsik, Y. Assemblability evaluation based on tolerance propagation. In Proceedings of the 1995 IEEE International Conference on Robotics and Automation, Nagoya, Japan, 21–27 May 1995.
8. Sanderson, A.C. Assemblability based on maximum likelihood configuration of tolerances. *IEEE Trans. Robot. Autom.* **1999**, *15*, 568–572. [CrossRef]
9. Cui, Z.; Du, F. Assessment of large-scale assembly coordination based on pose feasible space. *Int. J. Adv. Manuf. Technol.* **2019**, *104*, 4465–4474. [CrossRef]
10. Yuan, L.; Zhang, L.; Wang, Y. An optimal method of posture adjustment in aircraft fuselage joining assembly with engineering constraints. *Chin. J. Aeronaut.* **2017**, *30*, 222–229.
11. Wu, D.; Du, F. A Multi-constraints Based Pose Coordination Model for Large Volume Components Assembly. *Chin. J. Aeronaut.* **2019**. [CrossRef]
12. Ma, H.; Jin, Y.; Zhang, X.; Zhou, H. Complex shape product tolerance and accuracy control method for virtual assembly. *Proc. SPIE Int. Soc. Opt. Eng.* **2015**, *9446*, 94462E.
13. Du, Q.; Zhai, X.; Wen, Q. Study of the Ultimate Error of the Axis Tolerance Feature and Its Pose Decoupling Based on an Area Coordinate System. *Appl. Sci.* **2018**, *8*, 435. [CrossRef]
14. Davis, B.; Jones, T.M.; Darrell, D.Z.; Tracy, E. Digitally Designed Shims for Joining Parts of an Assembly. European Patents EP 2533167A2, 5 June 2012.
15. Fabian, S.; Nathapon, O.L.; Jorg, W. Automated assembly of large CFRP structures: Adaptive filling of joining gaps with additive manufacturing. In Proceedings of the 2016 IEEE International Symposium on Assembly and Manufacturing, Fort Worth, TX, USA, 21–22 August 2016; pp. 126–132.
16. Wang, Q.; Dou, Y.; Cheng, L.; Ke, Y.; Qiao, H.; Zhao, X. Shimming design and optimal selection for non-uniform gaps in wing assembly. *Assem. Autom.* **2017**, *37*, 471–482. [CrossRef]
17. Muelaner, J.; Kayani, A.; Martin, O.; Maropoulos, P.G. Measurement assisted assembly and the roadmap to part-to-part assembly. In Proceedings of the DET2011 7th International Conference on Digital Enterprise Technology, Athens, Greece, 27–30 September 2011; pp. 11–19.
18. Cui, Z.; Du, F.; Xiong, T. Analysis and coordination on assembly deviation of multi plane-and-holes assembly based on orientation points groups. *Acta Aeronaut. Astronaut. Sin.* **2018**, *38*, 248–257. (In Chinese)
19. Yang, J.L.; Huang, X.; Li, L.; Xiong, T.; Zhao, Z. Method for extracting repair amount of skin seam based on scan line point cloud. *Aeronaut. Manuf. Technol.* **2019**, *62*, 73–77. (In Chinese)
20. Yu, A.; Liu, Z.; Duan, G.; Tan, J.; Che, L.; Chen, X. Geometric design model and object scanning mode based virtual assembly and repair analysis. *Procedia CIRP* **2016**, *44*, 144–150. [CrossRef]
21. Manohar, K.; Hogan, T.; Buttrick, J.; Banerjee, A.G.; Kutz, J.N.; Brunton, S.L. Predicting shim gaps in aircraft assembly with machine learning and sparse sensing. *J. Manuf. Syst.* **2018**, *48*, 87–95. [CrossRef]
22. Lei, P.; Zheng, L. An automated insitu alignment approach for finish machining assembly interfaces of large-scale components. *Robot. Comput. Integr. Manuf.* **2017**, *46*, 130–143. [CrossRef]
23. Cheng, B. *Aircraft Manufacturing Coordination Accuracy and Tolerance Allocation*, 1st ed.; Aviation Industry Press: Beijing, China, 1987; pp. 1–16. (In Chinese)
24. Arun, K.S. Least-squares fitting of two 3-D point sets. *IEEE Trans. Pattern Anal. Mach. Intell.* **1987**, *9*, 698–700. [CrossRef]

25. Liu, J.; Wang, T.; Zou, C. A blending control aircraft assembly quality method using key assembly characteristic. *Adv. Mater. Res.* **2011**, *314–316*, 2469–2473.
26. Bourdet, P.; Mathieu, L.; Lartigue, C.; Ballu, A. The concept of small displacement torsor in metrology. *Adv. Math. Tools Metrol.* **1996**, *40*, 110–122.

Article

Deep Neural Network for Automatic Image Recognition of Engineering Diagrams

Dong-Yeol Yun [1], Seung-Kwon Seo [1], Umer Zahid [2,*] and Chul-Jin Lee [1,*]

[1] School of Chemical Engineering and Materials Science, Chung-Ang University, 84 Heukseok-ro, Dongjak-gu, Seoul 06974, Korea; dy7585@cau.ac.kr (D.-Y.Y.); sungkseo@cau.ac.kr (S.-K.S.)
[2] Chemical Engineering Department, King Fahd University of Petroleum and Minerals, Dhahran 31261, Saudi Arabia
* Correspondence: uzahid@kfupm.edu.sa (U.Z.); cjlee@cau.ac.kr (C.-J.L.); Tel.: +966-13-860-7360 (U.Z.); +82-2-820-5941 (C.-J.L.)

Received: 11 March 2020; Accepted: 4 June 2020; Published: 9 June 2020

Abstract: Piping and instrument diagrams (P&IDs) are a key component of the process industry; they contain information about the plant, including the instruments, lines, valves, and control logic. However, the complexity of these diagrams makes it difficult to extract the information automatically. In this study, we implement an object-detection method to recognize graphical symbols in P&IDs. The framework consists of three parts—region proposal, data annotation, and classification. Sequential image processing is applied as the region proposal step for P&IDs. After getting the proposed regions, the unsupervised learning methods, k-means, and deep adaptive clustering are implemented to decompose the detected dummy symbols and assign negative classes for them. By training a convolutional network, it becomes possible to classify the proposed regions and extract the symbolic information. The results indicate that the proposed framework delivers a superior symbol-recognition performance through dummy detection.

Keywords: convolutional neural network; object detection; piping and instrument diagram; unsupervised learning

---

1. Introduction

Engineering diagrams (EDs) are schematic drawings describing process flow, circuit construction, and engineering device information. Among the many types of EDs, piping and instrument diagrams (P&IDs) are broadly used in the production plant industry because they contain key information of the plant, including piping, valves, instruments, control logic, and annotations. Moreover, extracting this information—for example, where or of what type the objects are—is the first step in estimating the number of elements and managing the project during its operational period. Most of the plant industries, such as oil and gas production plants, have employed large teams of engineers to manually count these entities and digitalize the information into their internal systems because there is no module available to automatically extract such information from the diagrams. For decades, these tasks have been considered inefficient and time-consuming tasks. Consequently, the demand for a module enabling an automatic engineering diagram digitalization has increased as such procedures can improve productivity and gain a competitive edge for the company in the global market.

However, there are obstacles to be overcome before applying this technology in real-world scenarios. Firstly, the symbols of P&IDs come in a diverse range of forms, with approximately 100 different types for each entity [1]. Furthermore, there is an inter-similarity between these symbols themselves. This requires the person interpreting and counting the symbolic entities to know the P&ID symbols and legend sheets. In some diagrams, it is difficult to identify—through image processing alone—a target symbol without confusing it with another symbol because there are so many objects. Moreover, in diagrams, text information such as notes, as well as line number or size information, is presented near the symbols. This information is also often also written across the symbol; therefore, it can present another obstacle in the effective recognition of diagrams. These key challenges need to be overcome to enhance the capabilities of P&ID digitalization procedures in real-world scenarios; however, there have been few studies applied to develop an object-detection algorithm to overcome these limitations and present suitable applicability.

Within the field of machine vision for EDs, there have been a few studies that have sought to extract specific information from the diagrams. In [2], the authors present new trends on machine vision to extract various information from EDs, such as binarization, contextualization, segmentation, and recognition. One of the most popular preprocessing methods is binarization; by adopting a threshold value, this method converts an image into a binary representation, thereby removing noise and improving entity identification in the diagram. There are several methods of applying such image binarizations, including global thresholding [3], local thresholding, and adaptive thresholding [4]. For line detection in the diagrams, canny edge detection [5], Hough transformations [6], and morphological dilations have been discussed in the literature. Probabilistic Hough transform (PHT) [7], which uses a random sampling of the edge points to detect lines in images, has been applied for the robust detection of lines in engineering diagrams [8]. A shape-detection procedure, employing a consistency attributed graph (CAG) with a sliding window, was used by [9] to construct a symbol-detection procedure. As a comparable method, a recursive model of the morphological opening was implemented by [10] to identify symbols by the empty fraction of their area. For text/graphics segmentation (TGS), a connected component (CC) analysis [11] was used with size constraints in an engineering diagram [12]. In one study [13], a procedure to realize pixel-wise classification into text, graphics, and background was performed using filter banks and estimations of the descriptor sparseness. To find a specific shape in the diagrams, template matching was also used with a symbol shape incorporated as part of the prior information [14]. In [15], a threshold-based object-detection algorithm was proposed for binary images.

For graphical symbol recognition using a machine learning approach for engineering diagrams, there is very little previous research. In the past, the template matching method [16] has been famously used to find a specific shape within an image by sliding the template across the entire window. In all slides, the template calculated the similarity using convolutional estimation. However, this method has an inherent disadvantage; when it estimates the similarity within a specific region, it uses the Euclidian distance. This metric is intuitive, and the method is convenient to apply, but it cannot consider the images with large numbers of dimensions. It judges similarity using only quantitative calculations. Consequently, when it sets a high threshold value, it misses the shapes in the image owing to the stringent criteria. In contrast, when it sets a low threshold value, it detects unsuitable shapes because of its naïve criteria. This makes it difficult to find a suitable threshold value and improve its detection performance. Furthermore, almost all images in the real world contain noise or resolutions inadequate for the implementation of machine vision methods. In terms of industrial scenarios, the template matching-based method is not suitable for object detection in engineering diagrams.

In electrical EDs, circuit symbols can be recognized with morphological operations and geometric analyses [17]. A convolutional neural network can be used to recognize the symbols in hand-sketched engineering diagrams and convert them to a computer-aided design (CAD) program [18]. Adapting the Hopfield model, an iterative neural network model was implemented for symbol recognition by employing a prototype [19]. One symbol-classification method applied to P&IDs considered class decomposition using k-means clustering [20]. Fully convolutional networks (FCN), which are an end-to-end network for pixel-wise prediction, were first applied as object detection for P&IDs in [21]. In [22], the author proposed a method to extract various objects, including symbol, characters, lines, and tables in a P&ID, using a machine vision method containing deep learning architecture. To reduce human effort while validating the CAD documents, the P&IDs attributed by the graph form were trained by a neural network and predicted the components vector which represents the diagram flow in [23]. However, these previous approaches toward symbol recognition for P&IDs could not control the detection quality effectively for the symbols in the diagrams. Their applicability to real diagrams could not be confirmed because unpredictable objects were detected during the region proposal. This makes it difficult to recognize the target symbol owing to the characteristics of engineering diagrams made up of quasi-binary components.

In object detection, certain algorithms have delivered remarkable performances in recent years. The appearance of convolution networks has heralded significant improvements in image-classification problems. Even so, unlike basic image classification, object detection requires the solution not only of multi-labeled classification problems but also the bounding boxes for proposed regions in a digital image. To achieve this, networks for object detection employ a region proposal network (RPN), which plays the role of finding the symbol-region candidates, before classifying images. In a region-based convolutional neural network (R-CNN) [24], a selective search algorithm [25] was used for the RPN, and the boxed images were fed to the network for classification. RPN extracts numerous boxes from an image, considering the colors, scale, boundary, etc., of the object. The proposed regions are reshaped before being fed into the convolutional network for image classification. However, R-CNN has inherent limitations; it is expensive and slow. All processes in R-CNNs, from the RPN to the convolutional network, render the model inefficient and slow to detect objects in an image in real-time. Advanced models such as Fast R-CNN [26] and Faster R-CNN [27] improve network performance and speed by including the RPN in the neural network. In fast R-CNN, by employing a selective search algorithm as the RPN and implementing it into the neural network, it becomes possible to combine the different procedures into one end-to-end network. In Faster R-CNN, to reduce time consumption in the selective search algorithm, the algorithm was replaced by a combined neural network, which made the detection much faster. Many other state-of-the-art algorithms are being proposed as next-generation object identification strategies [28,29]. However, these have focused on problems such as the improvement of model accuracies for colored images, improvement of detection speeds, and image segmentation (i.e., the process of partitioning the image into a set of pixels with multiple segments). However, engineering diagrams have a characteristic difference from colored images; they are an almost-binary component matrix with a specific size and shape for each symbol. This makes it difficult for a model to classify symbols using only limited information.

As an advanced application of object detection in EDs, we propose an R-CNN architecture containing clustering. For the RPN, we implement a sequential image-processing method that is modified for two types of target symbols: valves and instruments. By modifying the image-processing method for region detection, we propose candidate symbol regions using size-based detection. After the RPN, we get the symbols, but we also get the meaningless regions inevitably, such as truncated line, curve, noise, and so on, we call these 'dummy'. The detected dummy images are decomposed by unsupervised learning methods, and negative classes are assigned to them for image classification. For images containing positive classes, which are our target symbols, the dataset is augmented with padding-block. Through a simple convolutional network, the multi-class classification model is trained and applied to new diagrams for the model test.

In this research, we propose a model based on an R-CNN architecture that features dummy image clustering. A sequential image-processing method is used for the RPN, instead of a selective search algorithm. After the RPN, the dataset is constructed by positive classes, and dummy clustering is applied to treat the unwelcomed detections as negative classes; thereby improving the classification performance of the convolutional network. The remainder of this paper is organized as follows. Section 2 provides the methodology for the extraction of target symbols from P&IDs. Based on our proposed method, the region proposal and symbol-recognition results are discussed in Section 3. Finally, we conclude the paper in Section 4.

2. Materials and Methods

Here, we propose our R-CNN framework for recognizing graphical symbols in P&IDs, as shown in Figure 1. There are two main types of graphical symbols targeted in this study: valves and instruments. These symbols have characteristics such as size and shape, as shown in Figure 2. Using these characteristics, we construct an RPN by modifying several image-processing techniques. There are two types of proposed regions: symbol and dummy. For data annotation of the symbols representing positive samples, a P&ID symbol and legend sheet are referred to, which provide the standard set of shapes and symbols for documenting the diagram.

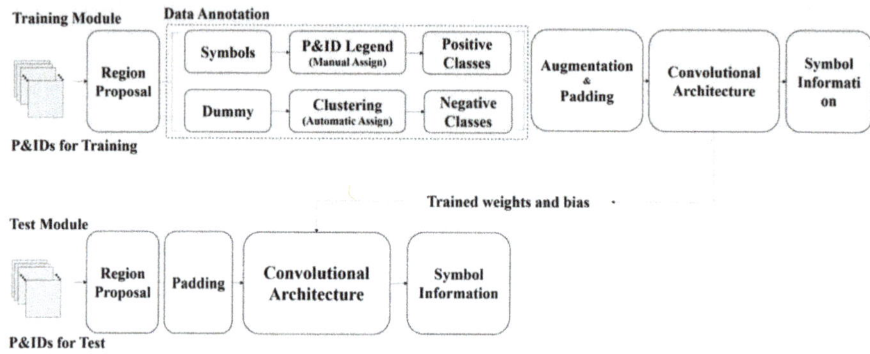

Figure 1. A framework summary for symbol recognition of the piping and instrument diagram (P&ID).

<table>
<tr><td colspan="2">Valve</td><td colspan="2">Instrument</td></tr>
<tr><td></td><td>Ball_Vlv</td><td></td><td>Sensor</td></tr>
<tr><td></td><td>Gate_Vlv</td><td></td><td></td></tr>
<tr><td></td><td>Globe_Vlv</td><td></td><td>PLC</td></tr>
<tr><td></td><td>3way_Vlv</td><td></td><td></td></tr>
<tr><td></td><td>Relief_Vlv</td><td></td><td></td></tr>
<tr><td></td><td>Butterfly_Vlv</td><td></td><td>Utility</td></tr>
<tr><td></td><td>Check_Vlv</td><td></td><td></td></tr>
</table>

Figure 2. Graphical characteristics of the target symbols.

For the dummy images representing negative samples, two unsupervised learning algorithms: k-means clustering and deep adaptive clustering (DAC), are used to analyze their hidden patterns and assign classes. After the annotation is completed, data augmentation is applied to generate additional information about the symbols. A convolutional neural network (CNN) is used to classify the symbols in this research owing to its superiority in local feature extraction [30]. After training the network, we apply it to another diagram in the same project and verify the results.

2.1. Data Sets

To implement the proposed framework and validate its performance, we use 10 pages of P&IDs from a real project. The resolution of the diagrams is 300 dpi in A3 size; thus, they contain approximately

4000 × 3000 cubic pixels. Of the 10 pages, we take seven and apply the region proposal method to construct our dataset of the proposed regions, which contains both positive and negative samples. After augmentation by padding (100 × 100 pixels), this is fed into a simple CNN.

We construct and compare three models based on the type of data that they use—positive samples only (P), positive with negative samples through k-means clustering (PN_Kmeans), and positive with negative samples through deep adaptive clustering (PN_DAC). To investigate and test our models, they are coded using Python with a Tensorflow backend. We also maintain the same computational conditions using NVIDIA TITAN V with 8 GB GDDR5.

2.2. Region Proposal

Region proposal is the process to extract candidate regions of symbols. Instead of apply the selective search algorithm, we build a customized process to extract the candidate regions in EDs as given in Figure 3.

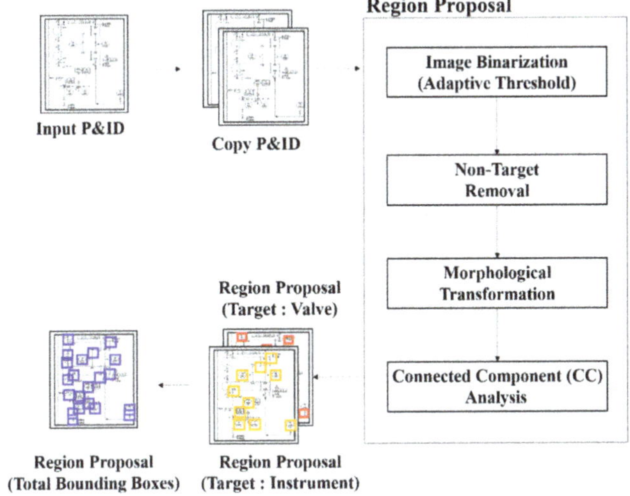

Figure 3. Procedure followed by the proposed region proposal network (RPN).

There are several points of incongruence in the selective search algorithm. For the proposal of candidate regions, the selective search algorithm uses the traits of an image, such as its color and boundary. However, this is not appropriate for the detection of an object in a binary image, such as an engineering diagram. To investigate the most-suitable algorithm for an engineering diagram, we implement sequential image processing to create proposal regions; this detects the potential target symbols by their characteristics. The sizes and shapes of plant symbols are specified in the diagram; therefore, it is possible to modify the image processing technique for each type of target symbol—valves or instruments—as they have a set size and aspect ratio in copies with identical resolutions. Thus, using copies of the input image, the characteristics of each target symbol can be reflected and sought for in each step. To modify the progress, we divide it into four parts: (1) image binarization, (2) non-target removal, (3) morphological transformation, and (4) CC analysis, as shown in Figure 3.

In image binarization, the adaptive threshold method [31] is used to reduce the noise present in the input images and convert them into a binary representation. Comparing a pixel against the average of those surrounding it preserves hard contrast lines and discards soft gradient changes.

In the non-target removal step, we remove the obstacles for the detection of each symbol. In the case of valve detection, other symbols such as lines, instruments, and pipe fittings are considered obstacles to clear detection. This process is employed as an intermediate stage to remove the obstacles and reduce the number of meaningless detections in the region-proposal step. For line removal, dilation kernels are used in the horizontal and vertical directions, with the structures being (1 × p) and (q × 1), respectively. The kernel parameters p and q are adjusted by considering the size of the symbol. In this study, we use a length similar to the shortest side of the symbol. For non-target symbol removal, a CC analysis algorithm [11] and Hough circle algorithm [32] are used to find the contours of the non-targeted objects such as instruments and pipe fittings.

In the morphological transformation step, "Closing," which is derived from the basic operations of erosion and dilation, is commonly used to enhance object outlines and small cover-up holes in the image [33]. Through this closing method, the floating objects retained from previous steps protect those background regions that have a similar shape to their kernel, while deleting all other background pixels [34].

Finally, to propose regions for the candidates of the target symbol, CC analysis is used with the constraints of symbol size and aspect ratio. The algorithm analyzes the topological structure of binary images. At the level of individual pixels, it considers 4-(8-) neighboring regions for the connected cases. We assume each symbol size and aspect ratio as prior knowledge in the detection. Figure 2 presents the schematic procedure of our region-proposal method. The image processing for region proposal plays a role in reducing the total number of proposed regions, and it adjusts the detection of undesirable objects called dummies.

After the region proposal step, we construct a dataset for the classification network as given in Figure 4. There are two types of proposed regions: symbol and dummy. Symbols are our positive samples; they are the gate valves, check valves, sensors, etc. Dummy entities have unpredictable shapes and sizes, are not within our interest, and make it difficult to classify symbols through the machine learning algorithm. Therefore, they are considered to be negative samples. For the positive dataset, we use the P&ID symbol and legend sheets, which provide a standard set of symbol shapes and legends for documenting diagrams and assign the symbol images for each class, such as gates, balls, globes, checks, etc. From the proposed regions, the positive samples are manually classified into 10 classes according to their shape and function.

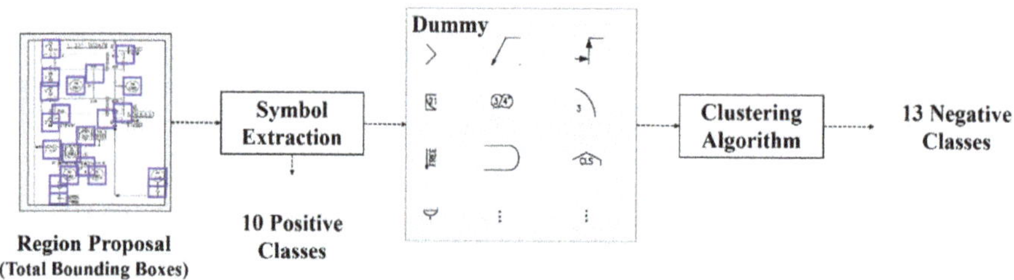

Figure 4. Data annotation: positive/negative samples.

2.3. Dummy Clustering

Aside from the positive samples, numerous images remain from the proposed regions, which are called dummies. They consist of curved lines, arrow shapes, revision clouds, or cut entities, as shown in Figure 3. It is difficult for the region proposal method to control the detection of dummy entities because the diagrams are quasi-binary representations and consist of many entangled lines and entities. In this research, we assign the negative classes of the detected dummies to improve the classification performance and consider the applicability of the procedure to real projects.

Dummies are obstacles for the classification model seeking to identify target symbols from a pool of proposed regions. P&ID is a type of grayscale image that is composed of only one channel, therefore, there is an arbitrary limitation to classifying the proposed regions when only using the positive samples from the target data. Furthermore, in the region-proposal network, the patterns of the detected dummies are unpredictable because it is difficult to erase all non-target entities during image processing. These patterns, such as shape, size, and detection frequency, are uncertain in every diagram, and this makes it difficult for the model to identify the target from the pool of detection images. Therefore, we assign negative samples to the classification models with unsupervised learning algorithms. To decompose the pool of dummy images and assign the class as a negative sample, we apply two unsupervised learning algorithms—k-means clustering and deep adaptive clustering (DAC). K-means clustering is a basic unsupervised learning algorithm. It is an iterative method to locating k-centroids in the dataset [35]; it locates them by optimizing the position of each centroid, based on the L2 norm in the feature space, as shown in Equation (1):

$$argmin_C \sum_{i=1}^{K} \sum_{x_j \in C_i} ||x_j - c_i||^2 \tag{1}$$

$$X = C_1 \cup C_2 \cdots \cup C_K, \; C_i \cap C_j = \phi \tag{2}$$

The quantity x represents a pool of unlabeled data, and c_i is a centroid of the i-th cluster, C_i.

DAC is also applied to decompose the hidden patterns of the dummies with an advanced method [36]. It is one of the state-of-the-art algorithms for the image-clustering problem that uses a convolutional architecture and cosine distance to measure the similarity of pairwise images with adaptive parameters. It delivers superior performance in image clustering owing to its adaptive-learning algorithm. The network solves the image-clustering problem as a binary pairwise-classification problem. The flowchart is presented in Figure 5.

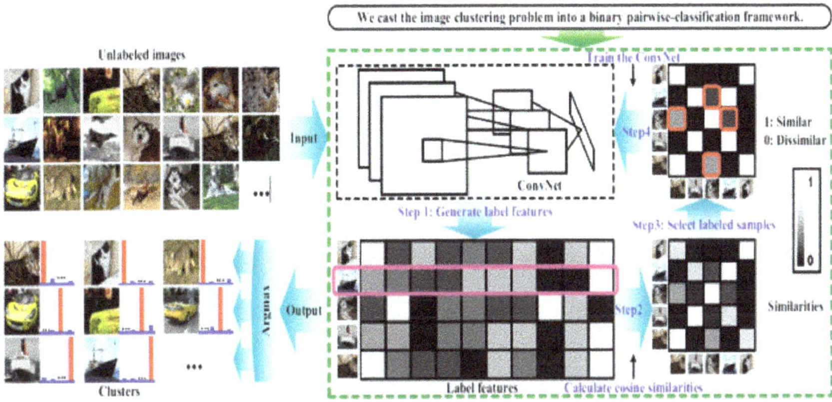

Figure 5. Flowchart of the deep adaptive clustering (DAC) algorithm [36].

Initially, unlabeled images are input to the convolutional network to generate a provisional latent vector for the image. Using the latent features, cosine similarities between pairwise images x_i and x_j are calculated; then, a confusion matrix is constructed for every batch. The network objective function is defined by:

$$Min_\theta E(\theta) = \sum_{i,j} L(r_{ij}, l_i \cdot l_j), \tag{3}$$

$$s.t. \; \forall i, \; ||l_i||^2 = 1, \; and \; l_{ih} \geq 0, \; h = 1, \cdots, k \tag{4}$$

where r_{ij} is the unknown binary variable; if the pair of input images are in the same cluster, then $r_{ij} = 1$, otherwise, $r_{ij} = 0$. $||\cdot||^2$ represents the L2 norm of a vector, and l_{ih} represents the h-th element of the label feature of the k-dimensional latent vector l_i. As the cosine similarity of the input image pair can be formulated by $l_i \cdot l_j$, the objective function of DAC is expressed by the loss between r_{ij} and $l_i \cdot l_j$. The expression for $L\left(r_{ij}, l_i \cdot l_j\right)$ is formulated as follows:

$$L\left(r_{ij}, l_i \cdot l_j\right) = -r_{ij} \log\left(l_i \cdot l_j\right) - \left(1 - r_{ij}\right) \log\left(1 - l_i \cdot l_j\right) \tag{5}$$

However, the unknown variable r_{ij} is prior information. Thus, an adaptive parameter λ is applied for the stepwise threshold value; we use $\mu(\lambda)$ and $l(\lambda)$ as the values for selecting similar (or dissimilar) image pairs.

$$r_{ij} = \begin{cases} 1, & if \; l_i \cdot l_j \geq \mu(\lambda) \\ 0, & if \; l_i \cdot l_j \leq l(\lambda), \; i, j = 1, \cdots, n \\ None, & otherwise, \end{cases} \tag{6}$$

In the clustering process, the value of λ starts at a specific value and gradually increases. Besides this, the relationships $\mu(\lambda) \propto \lambda$, $l(\lambda) \propto \lambda$, and $l(\lambda) \leq \mu(\lambda)$ are set in the algorithm. After finishing a batch process, the parameter λ is also updated by the gradient descent algorithm.

$$Min_\lambda E(\lambda) = \mu(\lambda) - l(\lambda) \tag{7}$$

$$\lambda := \lambda - \eta \cdot \frac{\partial E(\lambda)}{\partial \lambda} \tag{8}$$

Here, η represents the learning rate of λ. Using this adaptive modification of the parameter λ, the algorithm performs a stepwise selection between the pair images with increasing λ. The performance of the DAC is detailed for various datasets in [36]; it delivers the best performance in a binary image clustering problem, such as MNIST when compared against other clustering methods. Therefore, as an advanced method to decompose dummy images, DAC is applied in this research, and the results are compared with those of the k-means clustering.

The detailed architecture of DAC is summarized as follows. After the input images are padded by 100×100, we use six convolutional layers with a (3×3) kernel size, (1×1) stride, ReLU (Rectified Linear Unit) activation, and padding of the same structure in this network. The number of filters in each layer are 64, 64, 64, 128, 128, and 128, respectively. A max-pooling operation is applied with a (2×2) kernel and (2×2) stride. In fully connected layers, hidden units contain 128 and 64 nodes with ReLU activation. In all the layers, batch normalization is used to prevent the outputs of the hidden nodes from fluctuating. For adaptive learning, we set the selection-control equations according to Equations (9) and (10):

$$u(\lambda) = 0.95 - \lambda \tag{9}$$

$$l(\lambda) = 0.455 + 0.1 \cdot \lambda \tag{10}$$

There are 451 instances of dummy images from the seven pages of P&IDs. Using these two algorithms, the hidden patterns of the dummy pool are identified; then, we automatically assign them into k classes as negative samples. In this research, the value of k is fixed at 13. The optimum value of k is also an issue in clustering problems; however, we only focus on the effects of assigning a negative class for classification networks.

Table 1 presents the results of the data structure with positive and negative classes. Based on the clustering results, 13 negative classes are constructed, along with 10 positive classes. After annotating the data, which contains a total of 23 classes for the proposed regions, data augmentation [37] is applied to enhance the information in each class. To augment data, we apply two methods, central movement, and rotation. Central movement means that the extracted images from the RPN are padded by 100×100, before entering the network. We implement it moves $1 \times 1.3 \times 3$ pixels around the center of the image to catch a located symbol a little sideways for the same class. The rotation is also applied because some valves exist in a rotated form in the diagrams, so, in this study, only $45, 90, 135, 180, 225,$ $270,$ and 315 rotation angles were implemented to catch the rotated symbol.

Table 1. Data annotation and augmentation.

Annotation Type	Classes	Instances	Instances (after Augmentation)
Positive	10	1213	29,620
Negative	13	451	4610
Total	23	1664	34,230

2.4. Convolutional Network

Several machine-learning methods can be applied to the image-classification problem, including the support vector machine, random forest, and neural network-based models; however, we implement a simple convolutional neural network as our classification model to extract local information of the image data by convolutional and max-pooling filters [37].

The detailed model structure is presented in Figure 6. We construct three convolution layers and two fully connected layers in the network. The number of convolution filters in each layer is 64, 128, and 256, respectively. A kernel size of (3 × 3), a stride of (1 × 1), and a max-pooling layer with (2 × 2) filters are used for local feature extraction. Fully connected layers consist of 256 and 23 units, and the ReLUactivation function is used throughout our model, except for the end unit, wherein a softmax function is used. For generalization of the model, the dropout method [38] is applied, which is set at 0.7. The purpose of the dropout is to prevent overfitting problems in the neural network-based model by applying a zero forward-direction propagation value stochastically to every layer.

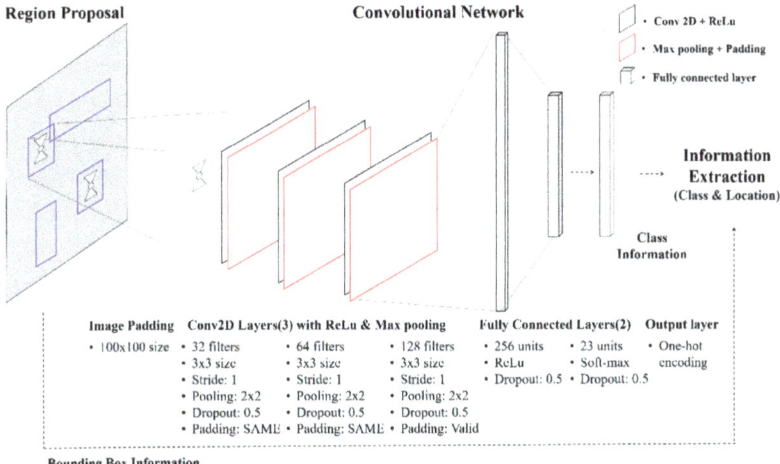

Figure 6. R-CNN scheme for symbol recognition of P&ID.

2.5. Evaluation Metric

Considering the target symbols in the diagrams and the requirements of practical applications, we suggest two metrics for validating the proposed framework-symbol recognition rate (SR) and dummy detection rate (DR), using a constant confidence threshold of 0.7.

$$SR\ (\%) = \frac{The\ Number\ of\ Correct\ Recognitions}{The\ Number\ of\ Symbols\ in\ the\ diagram} \times 100 \tag{11}$$

$$DR\ (\%) = \frac{The\ Number\ of\ Dummies\ Confused\ with\ Symbols}{The\ Number\ of\ Model\ Predictions} \times 100 \tag{12}$$

SR is the number of correctly recognized symbols divided by the number of symbols in the diagrams. It describes to what extent the model correctly detects the target symbols in the diagrams. *DR* is calculated by dividing the number of dummy images confused with symbols by the number of model predictions. It also describes the capacity of our model to distinguish dummies from symbols. If the model is well-trained in object detection for P&IDs, the value of *SR* will be large, whereas the value of *DR* will be small. In this study, the models are validated and compared with each other using these two metrics.

3. Results

3.1. Region Proposal Results

Figure 7 summarizes the region proposal results. The target symbols-valves and instruments were well-detected in these results. We implemented a customized procedure for each target symbol and integrated the proposed regions into one diagram. All the targets in the diagram were detected using image processing. For each target, the image processing was set to modify the overlapping contours in the detected regions.

Sine the proposed regions were detected by size constraints in the contour method—that is the CC analysis—there were unwelcomed images in the resulting diagram. These had a similar size to the target and represented sliced lines, the edges of instruments, entangled lines, etc. To reduce the number of dummy detections, the size constraints were used to customize the image processing for each target by adopting the target size as prior knowledge. The main purpose of the region proposal was to identify the candidate regions where the target symbol might exist; therefore, a noteworthy advantage of the process is that we are not required to focus on making the number of candidate symbols as small as possible; they must be detected conservatively and passed into the convolutional network for target identification.

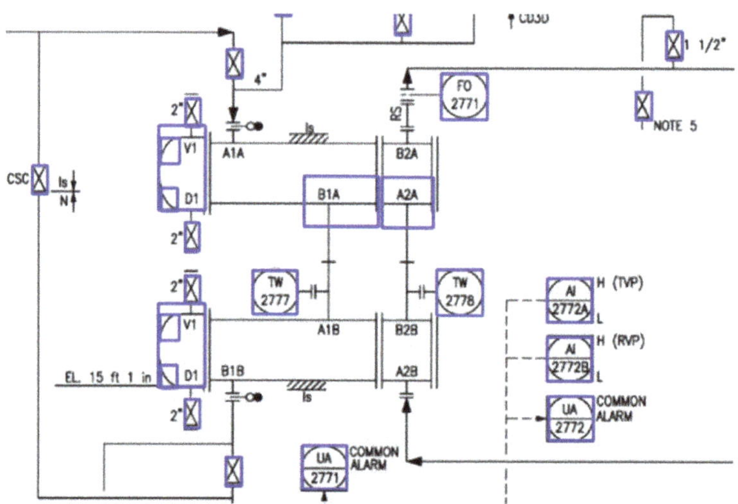

Figure 7. Sample results of the region proposal network (RPN).

From these proposed regions, we obtained a pool of images containing symbols and dummies. In the P model, only the symbol data are constructed as the dataset for the classification model. On the other hand, in the PN models, both symbols and dummies are incorporated into the model. To assign classes to the dummies, a series of detected dummies was decomposed through the clustering algorithms. Consequently, the effects of negative classes on symbol recognition in engineering diagrams were analyzed; these are described in the following section.

3.2. Effects of Negative Classes

First, we only investigated the positive samples to test the performance of the model. The model recognized symbols in the test diagrams but could not distinguish dummy images from the proposed region. This demonstrates that the model, which is only trained with positive samples, can distinguish only symbols. We could observe that both PN models-k-means and DAC were superior to the P model in terms of target-symbol classification from the proposed regions. In Figure 8a, the symbols were well recognized by the P model, but dummies were also detected in the results. This means that the model, which was trained only on positive data, had a weakness in identifying negative samples as false. In contrast, PN models such as Figure 8b exhibited strong discriminative performance between symbols and dummies. The dummies confused with the check valves and gate valves were filtered out by the PN models. These results indicated that the assignment of a negative class for classification gives the model the ability to effectively identify symbols from the pool of binary component images.

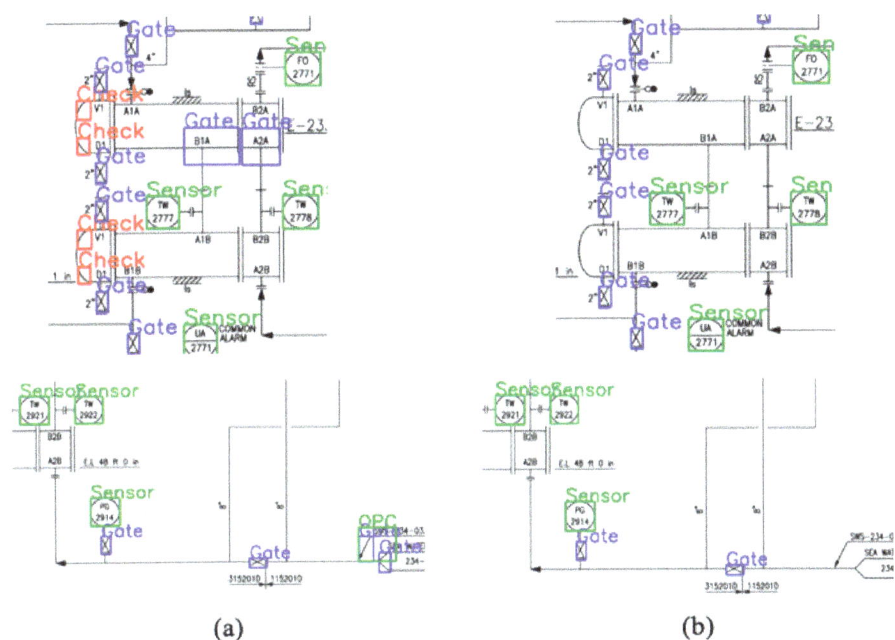

(a)   (b)

Figure 8. Sample results from the (**a**) P_model, (**b**) PN_Kmeans model.

We verified this statement in Table 2. In terms of SR—the extent to which the model could recognize the symbols in the diagram—all the models delivered good performance of over 96%. The PN_DAC model outperformed the other two, with 98.08%. This suggests that in the PN models, there was enhanced ability to classify targets through the assignment of a negative class.

Table 2. Results of the models.

Model Type	SR (%)	DR (%)
P Model	96.97	42.31
PN_Kmeans Model	97.88	1.35
PN_DAC Model	98.08	0.39

DR demonstrated the remarkable ability of both PN models. In the P model, 42.3% of the dummy images from the test diagrams were confused with our target symbols. This meant that the P model was incapable of identifying what images represented a genuine symbol. Due to the characteristics of EDs, i.e., their binary representation, it was difficult for the model to recognize them. In the latent space of the convolutional network, the latent features of both the dummies and the symbols were confused with each other under the P model because it does not possess any information concerning negative images. Hence, we conclude that negative classes are required for object-detection algorithms in EDs.

Compared to the P model, the models that consider negative samples achieve a significant reduction in the dummy detection rate. Binary images contain only a limited amount of information. Although most images in real-world applications consist of three channels-red, green, and blue-engineering diagrams consist of one channel-grayscale. They feature only one channel, which consists of quasi-binary components; hence, there is limited information available in the image, such as local features or pixel intensity. In this respect, we can say that for engineering diagrams to effectively recognize plant symbols and discard the detected dummies from the proposed regions, a dataset containing positive and negative classes is required. Consideration of the negative results yields the additional model information through which candidates can be assessed effectively.

3.3. Effect of Clustering Methods

In Table 3, it is shown in a confusion matrix to depict the performance of PN_DAC model.

The PN_DAC model exhibits the optimum performance in SR and DR, with 98.08% and 0.39%, respectively. Though both PN models-k-means and DAC-had a low dummy detection rate, the PN_DAC model recorded a lower score in dummy detection than the PN_Kmeans model by approximately 1%. This resulted from the differences between the image clustering methods. The k-means clustering is an iterative algorithm based upon Euclidian distance, which represents a simple quantitative distance between entities in feature space. It does not consider the direction of the feature. Consequently, the algorithm is too weak to construct with high-dimensional data such as that contained within image representations. In contrast, the PN_DAC model obtains the latent features of the data by performing efficient feature extraction using a convolutional network. As shown in Table 3, most of the confusion is created among these symbol classes, except for the three-way valves, ball valves, and sensor symbols. We also calculate F1 scores for each symbol in PN_DAC model, as given in Table 4. By solving the pairwise binary classification problem with adaptive parameters, the model delivered good performance that could be interpreted as a good analysis of the hidden patterns in the regions. DAC has configured the negative class to make it easier to distinguish between the classes, thereby increasing its performance by entering well-defined data into the model.

Table 3. Confusion matrix: PN_DAC model.

Actual	Prediction											Total
	3way_Vlv	Ball_Vlv	Gate_Vlv	Butterfly_Vlv	Check_Vlv	Relief_Vlv	Globe_Vlv	Utility	Sensor	PLC	Dummy	
3way_Vlv	4	0	1	0	0	0	0	0	0	0	0	5
Ball_Vlv	0	5	0	0	0	0	0	0	0	0	0	5
Gate_Vlv	0	0	214	0	0	0	0	0	0	0	5	219
Butterfly_Vlv	0	0	0	9	0	0	0	0	0	0	0	9
Check_Vlv	0	0	0	0	11	0	0	0	0	0	2	13
Relief_Vlv	0	0	0	0	0	3	0	0	0	0	1	4
Globe_Vlv	0	0	0	0	0	0	28	0	0	0	2	30
Utility	0	0	0	0	0	0	0	18	0	0	2	20
Sensor	0	0	0	0	0	0	0	0	190	0	0	190
PLC	0	0	0	0	0	0	0	0	0	21	2	23
Dummy	0	0	5	1	2	1	2	2	0	2	291	306

Table 4. F1 score for each class (PN_DAC).

Class	F1 Score
3way_Vlv	0.89
Ball_Vlv	1.00
Gate_Vlv	0.97
Butterfly_Vlv	0.95
Check_Vlv	0.85
Relief_Vlv	0.75
Globe_Vlv	0.93
Utility	0.90
Sensor	1.00
PLC	0.91

4. Conclusions

In this study, an R-CNN for engineering diagrams was proposed, taking negative classes into account. For an RPN, sequential image processing was modified for each target—valve and instruments. To annotate the negative class for the dummy images, two unsupervised learning algorithms–k-means and DAC–were applied to decompose the hidden patterns of the dummies, and assign negative classes. A simple convolutional network was used as the classification model because of its superior characteristics in terms of local information extraction from the images.

There were three types of datasets used for the classification problem—positive (P model), positive with negative through k-means (PN-Kmeans model), and positive with negative through DAC (PN-DAC model). Compared to the P model, both k-means and DAC had relatively low dummy detection rates of 1.35% and 0.39%, respectively, because the negative class from the unsupervised algorithm improved the model's ability to distinguish dummies from the symbols in the diagrams. Moreover, DAC was a superior algorithm for decomposing binary representations, as the PN-DAC model had superior performance, in which the symbol recognition rate (SR) and the dummy detection rate (DR) were 98.08% and 0.39 %, respectively.

From these results, we can verify that the proposed model meets the applicability and practicality criteria for P&ID object detection algorithms. The algorithm's negative sample detection reduces due to dummies, which makes its application to real projects difficult. This object detection algorithm is expected to contribute to the automatic digitalization of engineering diagrams. Regarding further work, state-of-the-art algorithms for object detection, such as Faster R-CNN, You Only Look Once(YOLO)_v3, and Single Shot Multi-Box Detector (SSD), could be modified to suit engineering diagrams. For real-world applications, a tiny-object detector also would be useful as a plant symbol recognition model.

Author Contributions: Conceptualization, D.-Y.Y.; methodology, D.-Y.Y.; software, D.-Y.Y. and S.-K.S.; validation, D.-Y.Y. and S.-K.S.; formal analysis, D.-Y.Y. and S.-K.S.; investigation, D.-Y.Y.; resources, D.-Y.Y.; data curation, D.-Y.Y.; writing—original draft preparation, D.-Y.Y. and S.-K.S.; writing—review and editing, D.-Y.Y. and S.-K.S.; visualization, D.-Y.Y.; supervision, C.-J.L. and U.Z.; project administration, C.-J.L. and U.Z.; funding acquisition, C.-J.L. All authors have read and agreed to the published version of the manuscript.

Funding: This research was supported by the Chung-Ang University Research Grants in 2018 and Seoul R&BD Program (20191471).

Conflicts of Interest: The authors declare no conflict of interest. The funders had no role in the design of the study; in the collection, analyses, or interpretation of data; in the writing of the manuscript, or in the decision to publish the results.

References

1. Howie, C.; Kunz, J.; Binford, T.; Chen, T.; Law, K. Computer interpretation of process and instrumentation drawings. *Adv. Eng. Softw.* **1998**, *29*, 563–570. [CrossRef]
2. Moreno-García, C.F.; Elyan, E.; Jayne, C. New trends on digitisation of complex engineering drawings. *Neural Comput. Appl.* **2018**, *31*, 1695–1712. [CrossRef]
3. Otsu, N. A threshold selection method from gray-level histograms. *IEEE Trans. Syst. Man, Cybern.* **1979**, *9*, 62–66. [CrossRef]
4. Sauvola, J.; Pietikäinen, M. Adaptive document image binarization. *Pattern Recognit.* **2000**, *33*, 225–236. [CrossRef]
5. Kittler, J.; Fu, K.S.; Pau, L.F. *Pattern Recognition Theory and Application*; D. Reidel Publishing Company: Dordrecht, Holland, 1981; Volume 81, pp. 292–305.
6. Ballard, D. Generalizing the Hough transform to detect arbitrary shapes. *Pattern Recognit.* **1981**, *13*, 111–122. [CrossRef]
7. Kiryati, N.; Eldar, Y.; Bruckstein, A. A probabilistic Hough transform. *Pattern Recognit.* **1991**, *24*, 303–316. [CrossRef]
8. Matas, J.; Galambos, C.; Kittler, J. Robust detection of lines using the progressive probabilistic hough transform. *Comput. Vis. Image Underst.* **2000**, *78*, 119–137. [CrossRef]
9. Yu, B. Automatic understanding of symbol-connected diagrams. In Proceedings of the 3rd International Conference on Document Analysis and Recognition, Montreal, QC, Canada, 14–16 August 1995; pp. 803–806.
10. Datta, R.; Mandal, P.D.S.; Chanda, B. Detection and identification of logic gates from document images using mathematical morphology. In Proceedings of the 2015 Fifth National Conference on Computer Vision, Pattern Recognition, Image Processing and Graphics (NCVPRIPG), Patna, India, 16–19 December 2015; pp. 1–4. [CrossRef]
11. Suzuki, S.; Abe, K. Topological structural analysis of digitized binary images by border following. *Comput. Vis. Graph. Image Process.* **1985**, *30*, 32–46. [CrossRef]
12. Fletcher, L.; Kasturi, R. A robust algorithm for text string separation from mixed text/graphics images. *IEEE Trans. Pattern Anal. Mach. Intell.* **1988**, *10*, 910–918. [CrossRef]
13. Cote, M.; Albu, A.B. Texture sparseness for pixel classification of business document images. *Int. J. Doc. Anal. Recognit. (IJDAR)* **2014**, *17*, 257–273. [CrossRef]
14. Mokhtarian, F.; Abbasi, S. Matching shapes with self-intersections: Application to leaf classification. *IEEE Trans. Image Process.* **2004**, *13*, 653–661. [CrossRef] [PubMed]
15. Tuncer, T.; Avci, E.; Çöteli, R. A new method for object detection from binary images. In Proceedings of the 2015 23nd Signal Processing and Communications Applications Conference (SIU), Malatya, Turkey, 16–19 May 2015; pp. 1725–1728.
16. Belongie, S.; Malik, J.; Puzicha, J. Shape matching and object recognition using shape contexts. *IEEE Trans. Pattern Anal. Mach. Intell.* **2002**, *24*, 509–522. [CrossRef]
17. De, P.; Mandal, S.; Bhowmick, P. Recognition of electrical symbols in document images using morphology and geometric analysis. In Proceedings of the 2011 International Conference on Image Information Processing, Shimla, India, 3–5 November 2011; pp. 1–6. [CrossRef]
18. Fu, L.; Kara, L.B. From engineering diagrams to engineering models: Visual recognition and applications. *Comput. Des.* **2011**, *43*, 278–292. [CrossRef]
19. Gellaboina, M.K.; Venkoparao, V.G. Graphic symbol recognition using auto associative neural network model. In Proceedings of the 2009 Seventh International Conference on Advances in Pattern Recognition, Kolkata, India, 4–6 February 2009; pp. 297–301.
20. Elyan, E.; García, C.F.M.; Jane, C. Symbols classification in engineering drawings. In Proceedings of the 2018 International Joint Conference on Neural Networks (IJCNN), Rio de Janeiro, Brazil, 8–13 July 2018.
21. Rahul, R.; Paliwal, S.; Sharma, M.; Vig, L. Automatic information extraction from piping and instrumentation diagrams. In Proceedings of the 8th International Conference on Pattern Recognition Applications and Methods, Prague, Czech Republic, 19–21 February 2019; pp. 163–172.
22. Yu, E.-S.; Cha, J.-M.; Lee, T.; Kim, J.; Mun, D. Features recognition from piping and instrumentation diagrams in image format using a deep learning network. *Energies* **2019**, *12*, 4425. [CrossRef]

Appl. Sci. **2020**, *10*, 4005

23. Rica, E.; Moreno-García, C.F.; Álvarez, S.; Serratosa, F. Reducing human effort in engineering drawing validation. *Comput. Ind.* **2020**, *117*, 103198. [CrossRef]
24. Girshick, R.; Donahue, J.; Darrell, T.; Malik, J. Rich feature hierarchies for accurate object detection and semantic segmentation. In Proceedings of the 2014 IEEE Conference on Computer Vision and Pattern Recognition, Columbus, OH, USA, 23–28 June 2014; pp. 580–587.
25. Uijlings, J.R.R.; Van De Sande, K.E.A.; Gevers, T.; Smeulders, A.W.M. Selective search for object recognition. *Int. J. Comput. Vis.* **2013**, *104*, 154–171. [CrossRef]
26. Girshick, R. Fast R-CNN. In Proceedings of the 2015 IEEE International Conference on Computer Vision (ICCV), Santiago, Chile, 11–18 December 2015; pp. 1440–1448.
27. Ren, S.; He, K.; Girshick, R.; Sun, J. Faster R-CNN: Towards Real-Time Object Detection With Region Proposal Networks. *IEEE Trans. Pattern Anal. Mach. Intell.* **2017**, *36*, 1137–1149. [CrossRef]
28. He, K.; Gkioxari, G.; Dollár, P.; Girshick, R. Mask R-CNN. In Proceedings of the 2017 IEEE International Conference on Computer Vision (ICCV), Venice, Italy, 22–29 October 2017; pp. 2980–2988.
29. Redmon, J.; Farhadi, A. YOLOv3: An Incremental Improvement. Available online: https://arxiv.org/abs/180402767 (accessed on 8 August 2019).
30. Krizhevsky, A.; Sutskever, I.; Hinton, G.E. ImageNet classification with deep convolutional neural networks. *Commun. ACM* **2017**, *60*, 84–90. [CrossRef]
31. Peuwnuan, K.; Woraratpanya, K.; Pasupa, K. Modified adaptive thresholding using integral image. In Proceedings of the 2016 13th International Joint Conference on Computer Science and Software Engineering (JCSSE), Khon Kaen, Thailand, 13–15 July 2016. [CrossRef]
32. Duda, R.O.; Hart, P.E. Use of the Hough transformation to detect lines and curves in pictures. *Commun. ACM* **1972**, *15*, 11–15. [CrossRef]
33. Gonzalez, R.C.; Woods, R.E.; Masters, B.R. Digital image processing, third edition. *J. Biomed. Opt.* **2009**, *14*, 029901. [CrossRef]
34. Vermon, D. *Machine Vision: Automated Visual Inspection and Robot Vision*; Prentice Hall: New York, NY, USA, 1994; Volume 30.
35. Vilalta, R.; Achari, M.-K.; Eick, C. Class decomposition via clustering: A new framework for low-variance classifiers. In Proceedings of the Third IEEE International Conference on Data Mining, Melbourne, FL, USA, 22–22 November 2003; pp. 673–676.
36. Chang, J.; Wang, L.; Meng, G.; Xiang, S.; Pan, C. Deep adaptive image clustering. In Proceedings of the 2017 IEEE International Conference on Computer Vision (ICCV), Venice, Italy, 22–29 October 2017.
37. Wiradarma, T.P. *Comparison of Image Classification Models on Varying Dataset Sizes*; Hasso Plattner Institute: Potsdam, Germany, 2015.
38. Hinton, G.E.; Srivastava, N.; Krizhevsky, A.; Sutskever, I.; Salakhutdinov, R.R. Improving Neural Networks by Preventing Co-Adaptation of Feature Detectors. 2012, pp. 1–18. Available online: http://arxiv.org/abs/1207.0580 (accessed on 12 August 2019).

 *applied sciences*

Article

Data-Driven Design Solution of a Mismatch Problem between the Specifications of the Multi-Function Console in a Jangbogo Class Submarine and the Anthropometric Dimensions of South Koreans Users

Jihwan Lee [1], Namwoo Cho [2], Myung Hwan Yun [2] and Yushin Lee [3],*

[1] Department of Technology Management Economic and Policy, Seoul National University, Seoul 08826, Korea; rhyjihwan@empal.com
[2] Industrial Engineering, Seoul National University, Seoul 08826, Korea; chonamwoo@snu.ac.kr (N.C.); mhy@snu.ac.kr (M.H.Y.)
[3] Institute of Engineering Research, Seoul National University, Seoul 08826, Korea
* Correspondence: keynote1112@gmail.com; Tel.: +82-10-2391-9199

Received: 26 November 2019; Accepted: 4 January 2020; Published: 6 January 2020

Abstract: The naval multi-function console provides various types of information to the operator. It is equipment that is key for submarine navigation, and fatal human errors can occur due to the mismatch between the console specifications and the operator's body size. This study proposes a method for deriving console specifications suitable for the body size of Korean users. The seat height, seat width, seat depth, upper edge of backrest, and worktable height were selected as the target design variables. Using six anthropometric dimensions, a mismatch equation for each target design variable was developed. Anthropometric measures of 2027 Korean males were obtained, and the optimal specifications of the console were derived via an algorithmic approach. As a result, the match rate, considering all the target design variables, was improved from 2.57% to 76.96%. In previous studies and standards, the optimal console specifications were suggested based on the anthropometric data of a specific percentile of users, and it was impossible to quantitatively confirm the suitability of the console design for the target users. However, the method used in this study calculated the match rate using the mismatch equation devised for comfortable use of the console and a large amount of anthropometric data that represented the user population, and therefore the improvement effect of the recommended specification can be directly identified when compared to the current specifications. Moreover, the methodology and results of this study could be used for deciding the specifications of multi-function consoles in several fields, including nuclear power plants or disaster situation rooms.

Keywords: multi-function console; data-driven design; mismatch equation; anthropometric measures; algorithmic approach; optimal design

---

1. Introduction

The naval multi-function console is part of the computer system of a battleship and it is designed for communication between the user and the computer. The console is connected to various sensors in the ship, it displays a variety of information, and the user is able to control the different types of information.

Although the crew members of South Korean Navy ships perform a variety of tasks depending on their position, most of the crew who are in the combat information and engine control rooms work in front of the console for more than 8 h on a daily basis. Console operators handle various types of information displayed on the console in a very concentrated state for a long period of time. Considering the working characteristics of the console operators, they could be affected by various musculoskeletal disorders such as turtle neck syndrome and carpal tunnel syndrome, as well as chronic diseases such as low back pain and neck pain, if the height of the worktable or seat is inappropriate for the user's body size [1]. In addition, the ongoing physical burden on the console operators could probably lead to unintended operational errors, thus, reducing the mission efficiency and dispersing the focus on console operations [2].

Anthropometry means measurements of the human body. It is derived from the Greek words anthropos (man) and metros (measure) [3], and is needed in the design of machines, tools, and work environments in order to improve well-being, health, comfort, and safety [4]. The anthropometric data widely influence furniture design, and thus workplace design since the matching of body dimensions and furniture dimensions is vital to promote proper body posture for the user. An absence of anthropometry consideration would, in most cases, result in uncomfortable design for the targeted users and worse, unsafe, and unhealthy conditions. Therefore, to make the workplace comfortable for a person it should be designed based on an individual user's anthropometric dimensions [5,6]. Because of the importance of anthropometry, many previous studies have applied anthropometric methodologies to the design of the workplace [7–11].

Considering the improper posture of the console operator and the resulting decrease in concentration, which may significantly impact the ability to conduct military operations, continuous efforts to find the right specifications for the console operator's body size are necessary. If human factors and ergonomics (HF&E) approaches are not considered in the multi-function console design, musculoskeletal disease and human errors are more likely to occur [12], and thus several studies have emphasized HF&E's importance in suggesting design guidelines for consoles [13–15]. ABS (2013), MIL-STD-1472G (2013), and NUREG-0700 (2003) issued in the United States, are widely used as standards to provide guidelines for maritime system design, military equipment design, and nuclear power plant facility design, respectively. However, these standards mainly focus on providing minimum requirements rather than optimal design parameters when HF&E departments have associated with designers and engineers. In addition, the suggested criteria have been set based on the anthropometric data of only U.S. citizens [12]. Moreover, the basis and procedure for the optimal specifications recommended by these standards are unclear, and it is difficult to clearly confirm the improvement effect of the proposed optimal specifications as compared with the existing specifications.

The Korean Navy has solely focused on software improvements for operational performance of the console, and little attention has been given to hardware improvements to create a comfortable and secure console operating environment for users. Additionally, in Korea, the research on the development of military products that reflect the characteristics of the user's body has been focused on combat support systems such as military winter clothes, combat suits, and boots, and there is a relative lack of research on the ergonomic design of weapon systems such as the multi-function console. In the case of a combat support system, it is possible to improve a part of the product or to change the product within a short period of time when it is introduced and used in the military. However, the application of new design methods in a weapon system requires a longer period of time for development, and much more attention should be paid to a user-centered environment than that of a combat support system, because such design should be used for more than 20 years.

In this study, the console operating environment of the Jangbogo class submarine is presented as an example of the problems that may occur in terms of ergonomics when the specifications of the console are inappropriate for the user's body size. The seat height of the Jangbogo class submarine can be adjusted vertically, but the lowest height is measured to be 475 mm; seat width, seat depth, upper edge of backrest, and worktable height of the Jangbogo class submarine are measured to be 490, 482, 510, 817 mm, respectively.

2. Methods

Figure 1 shows the procedure for evaluating and improving the specifications of the submarine's multi-function console.

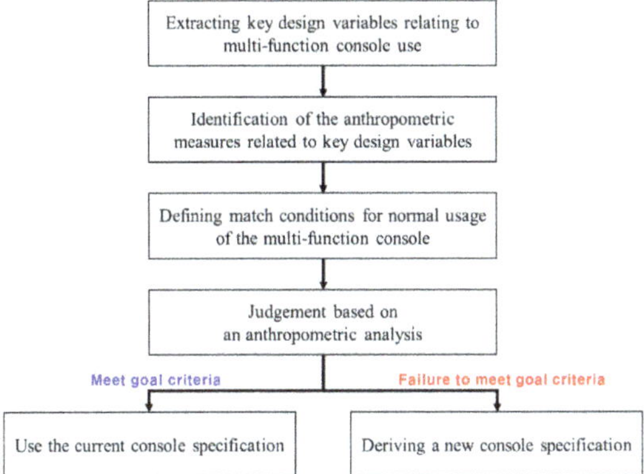

Figure 1. Procedure for design of submarine multi-function console using the anthropometric methodology.

First, the key design variables for multi-function console were extracted, and the detailed specifications of the current multi-function console were measured. Secondly, considering the context of the use of the console, anthropometric measurements related to the key design variables were identified. Third, the match conditions that guarantee a normal operation were also reviewed. Fourth, to judge the appropriateness of the current multi-function console specifications, the anthropometric dimensions were collected in consideration of the target population. Finally, after setting 70% of the match rate as the goal criteria, the suitability of the current multi-function console specifications was evaluated using a mismatch equation. In this study, an algorithmic approach was used to derive the optimal console specification, given that the match rate of the current multi-function console specifications did not reach the goal criterion.

2.1. Key Design Parameters for Naval Multi-Function Consoles

The multi-function consoles of Jangbogo class submarines have four consoles placed side-by-side, as shown in the first diagram in Figure 2. The four consoles are 2740 mm wide and 1300 mm high. The second diagram in Figure 2 is presented without the backrest to facilitate comprehension of the various design variables related to the seat. The third diagram in Figure 2 shows a lateral view of the console operator, also illustrating the specifications for different design variables.

First, the seat height (SH) of the seat refers to the vertical length from the floor to the highest portion of the seat pan. Previous studies related to the sitting posture at the work environment or to ergonomic design of student furniture have shown that the design of the SH is the utmost important factor. This means that determining the SH is the most important measure for solving a mismatch problem [16,17]. If the seat is too high, both feet are off the ground and high pressure is applied to the skin tissue behind the knee [18–20]. If the SH is too low, the seat pan does not support the thighs, and this can result in a large burden on the hips and an abnormally bent waist when sitting [21,22]. The current seat of the Jangbogo class submarine is designed to be adjustable for height. However, considering the fact that three or more console operators operate the console alternately in one day and a situation where the military is running an emergency training, there are instances when the operator has to switch quickly with the main console operator. Therefore, calculating the optimum height of the seat to return to the basic height would be very beneficial and effective in operating the console in terms of context of use.

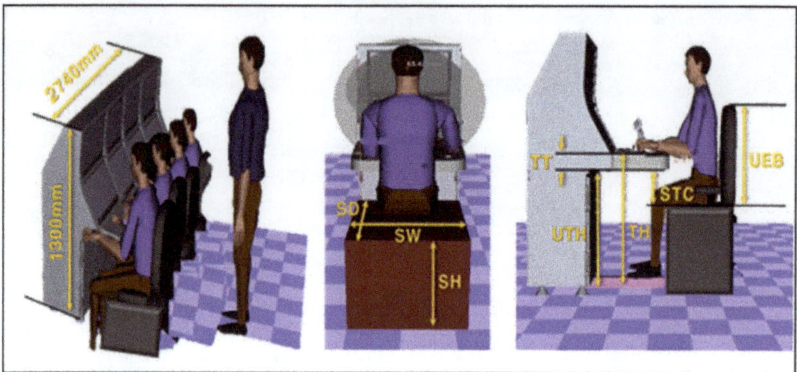

Figure 2. Submarine naval multi-function console design dimensions. Seat height (SH), vertical distance from the floor to the highest area of the seat pan; seat width (SW), horizontal distance from the left to the right side of the widest part of the seat pan; seat depth (SD), horizontal distance from the front to the rear of the longest part of the seat pan; upper edge height of backrest (UEB), vertical distance from the seat pan surface to the upper edge of the backrest (UEB); worktable height (TH), vertical distance from the floor to the worktable surface; underneath worktable height (UTH), vertical distance from the floor to the lowest point below the worktable; worktable thickness (TT), thickness of the worktable hardboard; and seat to table clearance (STC), vertical distance from the seat pan surface to underneath the worktable.

The seat width (SW) of the seat is the width from the left to the right side of the widest part of the seat pan. If the SW is too narrow, the sitting position may deviate from either side of the seat, and thus the width of the seat should be designed to be wider than the width of the user's hip [23–27]. Moreover, the upper limit of the SW needs to be considered, given that the seats are designed in a confined space and four console operators must sit side-by-side. In such context, Gouvali and Boudolos [28] argued that it is necessary to take into account an efficient utilization of the interior space in the submarine and to carefully derive the SW.

The seat depth (SD) is the length from the front to the back of the longest part of the seat pan. If the SD is too long, the backrest cannot support the back and waist properly, and the pressure between the front of the seat and the popliteal can increase, causing severe pain [20]. On the contrary, if the SD is too short, the pressure caused by the user's weight may not be evenly distributed through the user's hip and thigh, and the pressure may concentrate on a specific part of the body.

The upper edge height of backrest (UEB) means the vertical distance from the seat pan to the upper edge of the backrest. If the UEB is higher than the scapula, it may interfere with the free movement of the arms and torso [27,29]. Especially for console operators who work for more than 8 h per day, the above-mentioned situation can disable very basic activities such as stretching. However, if the UEB is too low, the back is not supported properly and this can induce excessive extension on the upper part of the back, which can lead to serious back injury.

The worktable height (TH) refers to vertical distance from the floor surface to the console platform surface. If the TH is too high, a console operator who frequently manipulates the keyboard and track ball installed on the worktable can suffer from excessive flexion and abduction of the shoulder and upper arm. In severe cases, this can lead to asymmetric spinal disorders. If the console TH is too low, the upper body is constantly bent forward and this can lead to kyphotic spinal posture [30].

The underneath worktable height (UTH) represents the vertical height from the floor to the lowest point of the worktable and the worktable thickness (TT) refers to the vertical distance from top to the bottom of the worktable.

The seat to worktable clearance (STC) represents the space between the seat and the worktable as the vertical distance from the extension of the seat pan surface to the bottom of the worktable.

This design variable is determined by the interrelationship between the seat and the worktable height. A too large STC implies that the seat is too low, or the worktable is too high. In such a case, discomfort can be induced in the shoulder and upper arm of the console operator, hindering normal shoulder movement. In contrast, if the STC is too narrow, sitting on the seat is not possible as the thigh can not enter between the seat and the worktable.

As the thickness of the worktable is fixed at 100 mm, there was no need for calculating the console UTH separately from the console TH. The STC between the seat and the worktable is also determined naturally when the SH and the TH are derived. The UTH, TT, and STC were measured to be 717, 100, and 142 mm, respectively.

Therefore, the SH, SW, SD, UEB, and TH were selected to be the final key design variables.

2.2. Anthropometric Criteria for Designing Naval Multi-Function Console in a Submarine

The age of the South Korean submarine crew ranges from 20 to 50 years, and they are only men. To consider the age of submariners, the seventh Korean anthropometric dataset for the age groups of 20–29, 30–39, 40–49, and 50–59 years were extracted from the survey made by SizeKorea in 2015. Six anthropometric measurements related to the target design variables of console operations were selected out of 133 anthropometric dimensions, as shown in Figure 3, which included: sitting thigh thickness (STT), popliteal height (PH), hip height (PH), hip width (HW), horizontal length between hips and ham (BPL), sitting shoulder height (SSH), and sitting elbow height (SEH).

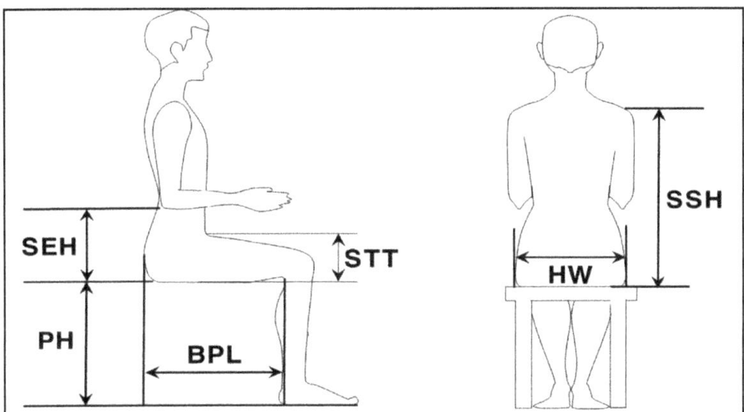

Figure 3. Anthropometric measures used in this study. Popliteal height (PH), vertical distance from the floor to the popliteal; hip width (HW), horizontal distance between the upper outer edges of the iliac crest bones of the pelvis; buttock to popliteal length (BPL), horizontal distance from the back of the buttocks to the popliteal; sitting thigh thickness (STT), vertical distance from the sitting surface to the superior thigh; sitting shoulder height (SSH), vertical distance from the sitting surface to the acromion; and sitting elbow height (SEH), vertical distance from the sitting surface to the underside of the elbow.

Descriptive statistical data of these six anthropometric measures for 2027 Korean males are presented in Table 1. They were used as the variables in the mismatch equation of this study.

Table 1. Anthropometric measures of Korean males between ages 20 and 50.

Anthropometric Measures	Mean (n = 2027)	SD	Min	Max	Percentiles		
					5	50	95
Popliteal height (mm)	428.01	20.5	353	523	395	428	463
Hip width (mm)	355.26	23.5	287	475	320	354	394
Buttock to popliteal length (mm)	490.46	22.9	420	592	454	490	530
Sitting thigh thickness (mm)	151.20	14.3	108	280	130	151	175
Sitting shoulder height (mm)	607.39	25.9	522	702	565	607	650
Sitting elbow height (mm)	268.79	25.1	195	364	227	270	309

2.3. Mismatch Equation for Naval Multi-Function Console in a Submarine

On the basis of the anthropometric measurements of the South Korean male, the mismatch equations used for specification of the key design variables in the submarine multi-function console define the maximum and minimum limits of those specifications.

All variables used in the mismatch equation were calculated in millimeter units.

First, the anthropometric dimensions for determining the SH was taken as the PH considering the sitting posture of the console operator as expressed in Equation (1). The shoe sole thickness (ST) was selected as the environmental variable.

$$(PH + ST) \times Cos30^{\circ} \leq SH \leq (PH + ST) \times Cos5^{\circ} \tag{1}$$

Equation (1) is based on constraints presented in Afzan, Hadi [31] and others [2,16,28,31–33], and this implies that the console operator should be able to extend at least 5° to 30° below their knees to feel comfortable when sitting in the seat. If the console operator sits at a right angle or at a smaller angle with the floor, fatigue can occur below the knee because of contraction of the tibial anterior muscle, and the excessive pressure can cause pain underneath the thigh if the knee is extended beyond 30°. The soles of submariners' shoes are designed to prevent onboard noise and shock and they are measured to be 40 mm thick. Equation (1) used this measure for the calculation.

The anthropometric variable used to determine SW was selected based on the body part in contact with the seat and the environment inside the submarine. As four consoles are arranged side-by-side, as indicated in Equation (2), HW and STT are adopted. The thickness of various control devices that are attached to the side of the seat pan, called manipulator thickness (MT), and the winter clothes thickness (WT) of console operators are used as environment variables.

$$HW < SW \leq 685 - [STT + MT + (WT \times 2)] \tag{2}$$

Equation (2) is based on the equations discussed by Castellucci and Arezes [16] and other research works [16,31,33], but these studies did not suggest an upper limit for the SW. In previous studies that observed settings in offices and schools, there usually was a huge clearance between seats, and the clearances did not cause excessive inconveniences or problems. A study by van Niekerk and Louw [34] and others even suggests that the SW should be designed from 1.1 times to 1.3 times the HW for the user's comfort and effective internal space utilization [28,32,34]. This study proposes an upper limit for SW considering the limited amount of space in a submarine setting, which requires it to be utilized in a very efficient manner. Figure 4 illustrates the deployment of four multi-functional consoles in Jangbogo class submarines.

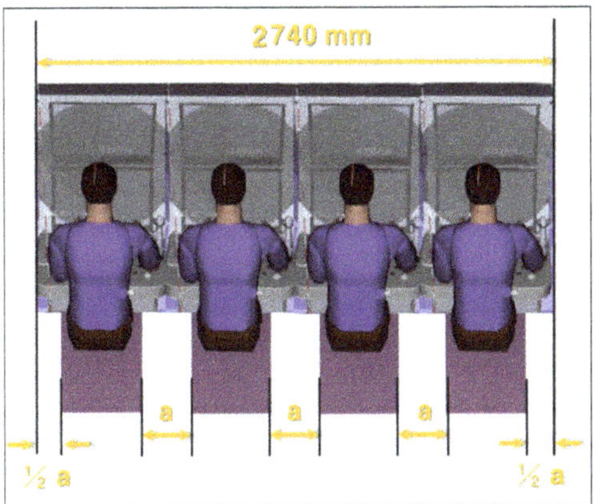

Figure 4. Deployment status of four submarine naval multi-function consoles.

In submarines, the consoles are arranged side-by-side to facilitate sharing of information among the four console operators. In the case of Jangbogo class submarines, only 2740 mm of horizontal space can be designed for all the consoles. If the seats are placed in the center of each console and if the distance between seats is represented by the character "a", the width of "a" should be at least wider than the thickness of the user's thigh considering the height of the seat because "a" should be designed to at least allow the console operators to enter and exit at "a". The fact that various control devices are installed on the side of the seat pan and the instance where the submariners are required to quickly return to the seat from working outside of the submarine to perform their tasks without taking off their thick winter clothes must also be taken into consideration.

Therefore, in this study, the thickness of the control device was fixed to be 20 mm and WT was fixed to be 10 mm resulting in a total thickness of 40 mm with the expression MT + (WT × 2).

Considering the sitting posture with user's back fully in contact with the backrest, the SD is determined using the BPL as shown in Equation (3) below.

$$0.80 \times BPL \leq SD \leq 0.95 \times BPL \tag{3}$$

Equation (3) was determined referencing to the equations used by Cotton and O'Connell [35] and others [2,16,31,33,35–43]. In particular, the coefficients presented in Equation (3) were calculated through various clinical trials in previous studies, and they were derived considering appropriate levels of comfortable knee extension and flexion when sitting with the hips and waist resting on the backrest.

As a parameter used for determining UEB, SSH was selected as shown in Equation (4) considering that the human body is in direct contact with backrest.

$$0.60 \times SSH \leq UEB \leq 0.80 \times SSH \tag{4}$$

Equation (4) was derived based on the findings of Agha [2] and other similar studies [2,31,32]. Each one of the coefficients, as those in Equation (3), was determined through a number of clinical trials. NUREG-0700 [15] recommends that the back of the seat should be able to support the lumbosacral region, which is the back curvature of the seat. Bendak and Al-Saleh [33] and Castellucci and Arezes [16] suggested only the upper limit of the UEB, stating that the UEB does not limit the basic upper body movement as long as the UEB is lower than the height of the user's subscapula. However, as emphasized in NUREG-0700 [15], if the UEB is low enough to fail to support the lumbar regions, it cannot properly support the back and waist, leading to their excessive extension. Therefore, the lower limit of the UEB must be also considered.

Equation (5) for the STC was devised based on the concept that the console operator's thigh should be able to fit under the worktable.

$$STT + 20 + (2 \times WT) \leq STC \tag{5}$$

The existing research recommended 20 mm for the sitting thigh thickness [28,29,36], but we added 10 mm considering the WT.

TH is determined by SH, thickness of the worktable, and STC. Therefore, this can be expressed in Equation (6) as follows:

$$TH = SH + TT + STC \tag{6}$$

Combining Equations (1), (5), and (6), the lower and upper limits of TH can be determined as expressed in Equation (7). In Equation (7), STT, SEH, PH, and SSH were selected as the anthropometric variables, whereas TT, WT, and ST were selected as environmental variables.

$$\mathrm{Max}[STT + 20 + (2 \times WT) + TT, \ SEH] + \left[(PH + ST) \times \mathrm{Cos}30^{\circ}\right] \leq TH$$
$$\leq (0.8517 \times SEH) + (0.1483 \times SSH) + \left[(PH + ST) \times \mathrm{Cos}5^{\circ}\right] \tag{7}$$

In Equation (7), the lower limit of the TH was chosen to be the higher value in between STC and SEH. This means that, in the sitting state, the height from the floor to the console operator's thigh should be lower than the UTH, and the elbow should be able to reach the worktable comfortably. If TH is lower than SEH, it would be very difficult to rest the elbows on the worktable without bending down, and manipulation of the keyboard and track ball would force the operator to bend forward. Considering the context of a console operator who heavily uses keyboards and track balls, the tension in the shoulder and back muscles can only increase if the elbows are not comfortably sitting on the worktable. The upper limit of the TH was derived by multiplying SEH and SSH by specific coefficients and then adding the calculated numbers to the upper limit of SH. The coefficients multiplied by SEH and SSH are given by the research of Parcells and Stommel [36] and Chaffin [44], who mathematically calculated the range of motion of the shoulder's flexion and abduction when working on a worktable and resting the arms on the worktable. If the height of the worktable is greater than the upper limit suggested by Equation (7), the shoulders can be excessively elevated upwards, or the arms are opened too widely to the sides when the elbows are raised on the worktable. This can cause increased fatigue and lead to musculoskeletal disorders of the shoulder and arm after a period of repeated tasks with the given environment.

2.4. Data Treatment

The minimum and maximum acceptable limits were calculated using the mismatch equation with specifications of six anthropometric measures. The equation was substituted with the anthropometric measurements of 2027 Korean males in the age groups of 20 to 29, 30 to 39, 40 to 49, and 50 to 59 years to verify whether the current multi-function console is suitable for the Korean body sizes. Each design specification of the current console that mismatched the Korean anthropometric dimension was determined, and the reasons behind the mismatch were analyzed. This study used Excel 2016 and SPSS25.0 to analyze the data. In addition, the greedy algorithm approach, which was utilized by Lee and Kim [45] to find the optimal height system for the chairs and desks of Korean students, was applied to derive the optimal specifications for the target design variables, and R programming was used to implement the greedy algorithm to calculate the optimal specifications of the console. The greedy algorithm approach is simple and primitive as it finds the maximum match rate of a specification by substituting the anthropometric dimension of each user in the mismatch equation and incrementing it by 1 mm sequentially for all possible specifications. Despite the simplicity of this algorithm, so far, it is essentially the best possible polynomial time approximation algorithm for the maximum coverage problem [46].

3. Results

Figure 5 shows the mismatch rates of the key design variables of the current multi-function console.

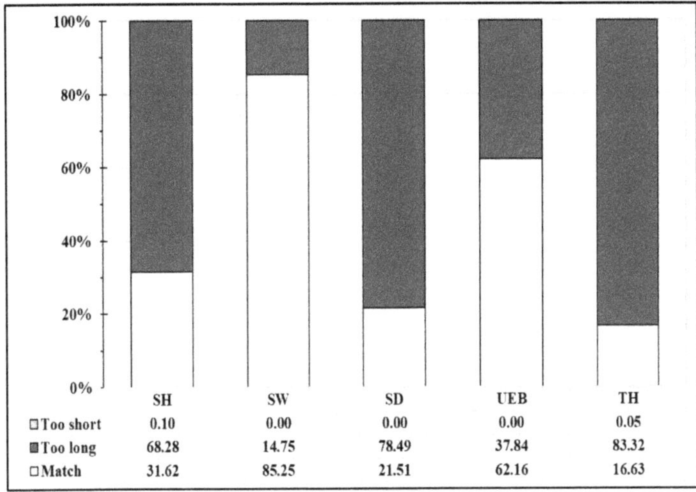

Figure 5. Mismatch rate of design specifications of the present submarine console.

First, the match rate of the current SH of Korean male body size was found to be 31.62%. In particular, the current SH was found to be higher than the body size of most men (68.28% of total), who were determined to be mismatched for the current size of SH. This means that the current SH is excessively high considering the user's PH. In this case, the majority of men were incapable of naturally touching the floor with their feet while resting their back on the backrest.

Secondly, the match rate of the current SW to the Korean male body size was 85.25%, and SW was considered wider than the HW of all men. It was found that 14.75% of men were mismatched for the current SW, which is wider than their anthropometric dimensions. The upper limit of the SW proposed in Equation (2) was defined only to set an effective utilization of the limited space in a submarine and, given that 14.75% of the Korean anthropometric dimensions were determined to be mismatched, the current SW do not present any problem for sitting purposes.

Third, the match rate of the current SD turned out to be 21.51%. In particular, the current SD was identified as inadequate for 78.49% of men as it was too long for their anthropometric dimensions. This indicated that the current SD is relatively longer than the BPL. Therefore, these men cannot sit with their backs in contact with the backrest or cannot bend their knees while sitting down. They are very likely to sit very unnaturally or uncomfortably, for instance sitting on the end of the seat while operating the console.

Fourth, the match rate of the current UEB for the Korean male size was 62.16%, and it showed the highest match rate of all the key design variables. However, 37.84% of men were identified to be mismatched for the current backrest height, which is higher than their scapula height. By limiting their upper body rotation and basic movements, the current backrest height can stiffen the user when operating the console for a long time.

4. Discussion

4.1. Analysis of Mismatch Conditions

This section examines whether the specifications of the multi-function consoles currently installed in Jangbogo class submarines meet the specifications recommended in previous studies or the standards. The result of the mismatch equation is also carefully analyzed and organized for each key design variable.

First, the current SH was found to be too high for the body size of the majority of Korean males. ABS [13] and MIL-STD-1472G [14] recommend that the SH is between 380 to 540 mm considering the user's PH. Although the current SH complies with the range suggested by the above-mentioned standard, considering the ST (40 mm), the current SH should be lowered as the average PH of Korean males is 428.01 mm, and sitting with 475 mm of SH can cause discomfort to the users as their feet do not touch the floor.

Secondly, the match rate of the current SW was 85.25%, which was significantly higher than the match rate of the other target design variables. MIL-STD-1472G [14] and ISO9241-5 [47] recommended that the seat should be at least 460 mm wide to fit the person with the widest hip. The current SW was 490 mm, and thus it was confirmed to meet the recommended specification. In addition, considering the fact that the size of the widest HP of Korean male is 475 mm, the current SW is not expected to cause any difficulty to the sitting task of console operators. However, the current seat is too wide for 14.75% of men and the SW could be narrowed to more effectively utilize the limited space in the submarine. It would not be a big problem to make the SW slightly narrower than it is now.

Third, the current SD has a match of only 21.51% for the Korean male body size and it turned out to be the worst fit for most men. NUREG-0700 [15] and MIL-STD-1472G [14] recommended that the depth of the seat should be from 381 to 431.8 mm considering the body size with the shortest BPL. The current SD of the seats installed in Jangbogo class submarines is 482 mm, and thus it is much longer than what is recommended. Among the Korean male anthropometric dimensions used in this study, the dimension of the user with the shortest BPL is only 420 mm and the BPL of users in the fifth percentile is only 454 mm. Therefore, it would be very difficult for them to bend their knees comfortably while leaning back on their backrest and sitting with a correct posture on the seat. There is a need to improve the SD by reducing the depth.

Fourth, the match rate of the current UEB was 62.16% and it is considered higher than the match rate of other key design variables. MIL-STD-1472G [14] recommended that the UEB should be from 480 to 580 mm so that users can support their torso well while they are sitting. The current UEB of the Jangbogo class submarine seat is 510 mm and it is considered to be in the recommended range. However, 37.84% of users have a high UEB, and hence they are hindered from making basic upper body movements. In addition, considering the unique usage context of the submarine console, where there is an administrator who monitors the console information from behind the seat, it is necessary to lower the UEB of the current seat.

Finally, the match rate of the current TH to the Korean male body size was only 16.63%. The TH is closely related to the SH, STT, and PH [47]. MIL-STD-1472G [14] and ABS [13] recommended that the TH should be in the range 740–790 mm and 650–810 mm, respectively. However, the current TH in the Jangbogo class submarine is 817 mm, which is greater than the height recommended by the standard. When the user works on a worktable that is higher than his or her body size, the manipulation of the keyboard and track ball tasks for a long period of time can be restricted because the comfortable operation of the shoulder joint and upper arm is not guaranteed.

Meanwhile, the STC is naturally determined by the SH and the TH. ISO9241-5 [47] suggested that the STC should be designed in consideration of human body size with the thickest thigh, and NUREG-0700 [15] recommended the STC to be at least 190.5 mm. The STC of the Jangbogo class submarine is currently 242 mm, which satisfies the recommended specification of NUREG-0700 [15]. However, considering that the thickest STT measurement from SizeKorea is 280 mm, the vertical adjustable range of the seat should be lowered further downwards.

The ISUS 83 combat command system and multi-function console of the Jangbogo class submarine were acquired from Germany in 1992, and these were developed in the early 1980s to enhance the performance of the German Navy's 206 submarine. Therefore, it is very likely that these consoles were built reflecting the dimensions of the German human body size measured in the 1980s. The German adult male had an average height of 180.5 cm in 1980 [48], whereas the Korean average height was only 172.9 cm in 2015. Therefore, it is natural that the size of the console designed for the German body size at the time mismatched the Korean male's anthropometric dimensions. Therefore, to obtain the optimal design specifications for the console matching the Korean anthropometric dimensions, the specification for each key design variable is proposed in Section 4.2, based on the results of the above analysis.

4.2. Recommendations for the Specifications of Submarine Naval Multi-Function Consoles Considering South Korean Body Size

One of the most commonly used methods in the development of standard systems, which was used in previous studies determining specifications of furniture for students, is the Ellipse methodology [17,49,50]. This method recommends an appropriate design range based on the fifth to 95th percentile dimensions of the collected anthropometric dimensions. For example, this method is implemented when determining the size of a hat; the head circumferences of the fifth and 95th percentile hat users are measured, and then, the size of the hat is determined within the range of the two.

In the case of the console, there are more anthropometric considerations to determine the specifications of each key design variable. To produce a single specification that can accommodate as many users as possible, it would be more appropriate to search for the optimal specification with the maximum coverage problem rather than the elliptic methodology.

To maximize match rate between each specification of the key design variables and anthropometric dimensions, the specifications listed in Table 2 were found to be the optimal.

Table 2. Recommended specifications for the South Korean submarine console.

Design Variable	SH	SW	SD	UEB	TH
Recommended Specification	431 mm	442 mm	429 mm	442 mm	738 mm

Among the recommended specifications presented in Table 2, SH, SD, and TH were within the recommend ranges in the previous standard, and UEB was approximately 38 mm lower than the existing standard. However, considering that the previous standards are from measurements in the United States and the fact that UEB is lower than the previous standard, while all the other design variables meet the recommended specification at the lower limit, it is inferred that the recommended specifications in Table 2 more practically reflect the Korean anthropometric dimensions for submarine consoles than the previous standards.

The existing standard recommends the SW to be wider than 460 mm, while the derived specification from the algorithmic approach was narrower by 18 mm. The widest hip width (475 mm) of Korean male adults cannot sit in the recommended SW but it is enough to fit the 95th percentile (394 mm). Considering the limited space inside the submarine, the seat specifications are considered to be appropriate. In addition, according to the recommendation in Table 2, the STC is 207 mm, which meets the minimum recommended standard proposed by NUREG-0700 [15]. Given that the seat can be adjusted vertically, when the SH is adjusted at a lower level, the console operators with the thickest thigh will be able to use the console.

Figure 6 shows a comparison between the match rate of the current console specification and that of the recommended specification.

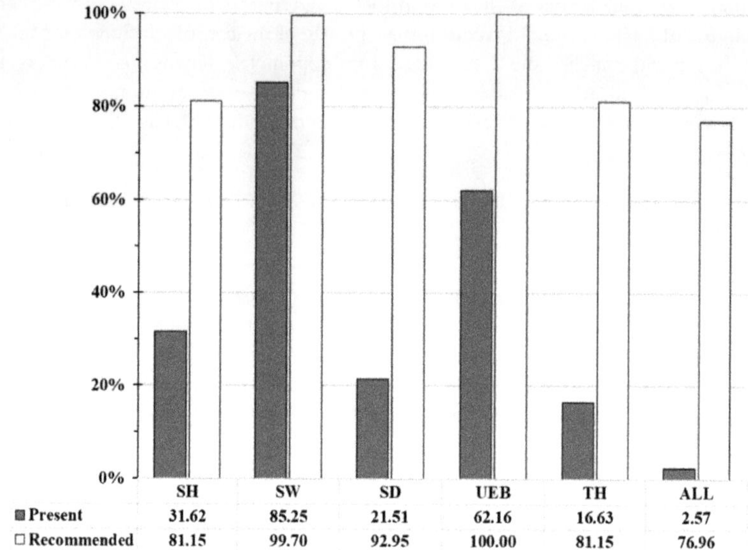

Figure 6. Comparison of match rate between the present and recommended specifications of the submarine console.

The match rates of the SH and TH are 81.15%, and it has been found that many more users can be accommodated than before. The SW, SD, and UEB are expected to fit 99.70%, 92.95%, and 100% of the Korean body sizes, respectively.

The rate of Korean males who have a suitable human body size for all five target design parameter specifications in the current console specification is only 2.57%, thus, on the one hand, 97.43% of users may have difficulties in using current console. On the other hand, it is expected that 76.96% of users have a suitable human body size for recommended console specification, and therefore many more users should be able to use the console comfortably as compared with the previous current console. Furthermore, considering the SH is adjustable, the number of the users who can comfortably use the console designed according to the recommended specifications is expected to be much higher.

In this study, human body size of only Korean males was used in calculating optimal design specifications of submarine console used by Korean submariners. However, if the anthropometric data for American or German users is applied with the methodology used in this study, it is expected that the optimal console specifications suitable for Americans or Germans could also be easily derived.

5. Conclusions

In this study, we derived the optimal design specifications for a multi-function console of Jangbogo class submarines that can accommodate, as much as possible, the anthropometric dimensions of Korean males.

To calculate the appropriate ranges for the key design variables, the working posture, the working environment, and the cooperation situation with other operational personnel were considered. The anthropometric dimensions of 2027 Korean male adults were substituted in the mismatch equation of each design variable to confirm the suitability of the Korean male body size for the current console.

All the key design variables, except the SW, were found to be inappropriate for the majority of Korean male's body sizes. To solve these problems of mismatch, we derived the optimal console specification through an algorithmic approach. As a result of calculating the match rate, it was found that the match rate can be improved up from 2.57% to 76.96% if the console is designed with the specifications proposed in this study.

The mismatch equation and algorithmic approach used in this study could further be used as a guideline for the specification of various military consoles in Korea, and it could be used to design work environments in various fields that operate multi-function consoles such as nuclear power plants and disaster control centers.

However, the mismatch equation used in this study is based on the previous studies dealing with the optimization of school furniture for the students or children. Therefore, it is necessary to verify through further empirical experiments whether the proposed mismatch equation is also valid for the working environment of the multi-function console and to continuously improve the equation if needed. In addition, this study was limited to the search for the optimal specifications of design variables related to the height of submarine consoles. Thus, in future research, the optimal specifications of distance-related design variables, such as depth of worktable, horizontal distance between seats and worktables, and placement radius of the various control buttons on the console, should be explored based on the reach envelope of Korean users.

Author Contributions: Conceptualization, J.L. and Y.L.; methodology, J.L.; software, Y.L.; validation, N.C., M.H.Y.; formal analysis, J.L. and Y.L.; investigation, M.H.Y.; resources, J.L.; data curation, N.C.; writing—original draft preparation, J.L.; writing—review and editing, Y.L.; visualization, N.C.; supervision, Y.L. All authors have read and agreed to the published version of the manuscript.

Funding: This research received no external funding.

Conflicts of Interest: The authors declare no conflict of interest.

References

1. Castellucci, H.; Arezes, P.; Molenbroek, J. Equations for defining the mismatch between students and school furniture: A systematic review. *Int. J. Ind. Ergon.* **2015**, *48*, 117–126. [CrossRef]
2. Agha, S.R. School furniture match to students' anthropometry in the Gaza Strip. *Ergonomics* **2010**, *53*, 344–354. [CrossRef]
3. Bridger, R. *Introduction to Ergonomics*; CRC Press: Boca Raton, FL, USA, 2008.
4. Pheasant, S.; Haslegrave, C.M. *Bodyspace: Anthropometry, Ergonomics and the Design of Work*; CRC Press: Boca Raton, FL, USA, 2018.
5. Bhuiyan, T.H.; Hossain, M.S.J. University hall furniture design based on anthropometry: An artificial neural network approach. *Int. J. Ind. Syst. Eng.* **2015**, *20*, 469–482. [CrossRef]
6. Sutalaksana, I.Z.; Widyanti, A. Anthropometry approach in workplace redesign in Indonesian Sundanese roof tile industries. *Int. J. Ind. Ergon.* **2016**, *53*, 299–305. [CrossRef]
7. Wang, E.M.-Y.; Wang, M.-J.; Yeh, W.; Shih, Y.-C.; Lin, Y.-C. Development of anthropometric work environment for Taiwanese workers. *Int. J. Ind. Ergon.* **1999**, *23*, 3–8. [CrossRef]
8. Reis, P.F.; Peres, L.S.; Tirloni, A.S.; dos Reis, D.C.; Estrázulas, J.A.; Rossato, M.; Moro, A.R. Influence of anthropometry on meat-packing plant workers: An approach to the shoulder joint. *Work* **2012**, *41* (Suppl. 1), 4612–4617. [CrossRef]
9. Hsiao, H.; Whitestone, J.; Bradtmiller, B.; Whisler, R.; Zwiener, J.; Lafferty, C.; Kau, T.Y.; Gross, M. Anthropometric criteria for the design of tractor cabs and protection frames. *Ergonomics* **2005**, *48*, 323–353. [CrossRef]
10. Ghaderi, E.; Maleki, A.; Dianat, I. Design of combine harvester seat based on anthropometric data of Iranian operators. *Int. J. Ind. Ergon.* **2014**, *44*, 810–816. [CrossRef]
11. Chang, C.-C.; Robertson, M.M.; McGorry, R.W. Investigating the effect of tool design in a utility cover removal operation. *Int. J. Ind. Ergon.* **2003**, *32*, 81–92. [CrossRef]
12. Rhie, Y.L.; Kim, Y.M.; Ahn, M.; Yun, M.H. Design specifications for Multi-Function Consoles for use in submarines using anthropometric data of South Koreans. *Int. J. Ind. Ergon.* **2017**, *59*, 8–19. [CrossRef]
13. ABS. *Guidance Notes on the Application of Ergonomics to Marine Systems*; American Bureau of Shipping: Houston, TX, USA, 2013.
14. *MIL-STD-1472G*; Department of Defense Design Criteria Standard Human Engineering: Washington, DC, USA, 2012.

15. *Human-System Interface Design Review Guidelines (Rev.2)*; NUREG-0700; Nuclear Regulatory Commission: Washington, DC, USA, 2003.

16. Castellucci, H.; Arezes, P.; Viviani, C. Mismatch between classroom furniture and anthropometric measures in Chilean schools. *Appl. Ergon.* **2010**, *41*, 563–568. [CrossRef] [PubMed]

17. Ymt, K.-R. Revision of the design of a standard for the dimensions of school furniture. *Ergonomics* **2003**, *46*, 681–694.

18. De Biomecánica Ocupacional, G.; Page, Á.; Molina, C.G. *Guía de Recomendaciones para el Diseño de Mobiliario Ergonómico*; Instituto de Biomédica de Valencia: Valencia, Spain, 1992.

19. Gutiérrez, M.; Morgado, P. *Guía de Recomendaciones para el Diseño del Mobiliario Escolar Chile*; Ministerio de Educación and UNESCO: Santiago, Chile, 2001.

20. Milanese, S.; Grimmer, K. School furniture and the user population: An anthropometric perspective. *Ergonomics* **2004**, *47*, 416–426. [CrossRef] [PubMed]

21. Cox, L. *Anthropometrics: An Introduction for Schools and Colleges*; Pheasant, S.T., Ed.; British Standards Institution: London, UK, 1984.

22. Knight, G.; Noyes, J. Children's behaviour and the design of school furniture. *Ergonomics* **1999**, *42*, 747–760. [CrossRef] [PubMed]

23. Evans, W.; Courtney, A.; Fok, K. The design of school furniture for Hong Kong schoolchildren: An anthropometric case study. *Appl. Ergon.* **1988**, *19*, 122–134. [CrossRef]

24. Occhipinti, E.; Colombini, D.; Molteni, G.; Grieco, A. Criteria for the ergonomic evaluation of work chairs. *La Med. Del Lav.* **1993**, *84*, 274–285.

25. Orborne, D. *Ergonomics at Work: Human Factors in Design and Development*; John Wiley and Sons: Chihester, UK, 1996.

26. Helander, M. Anthropometry in workstation design. In *A Guide to the Ergonomics of Manufacturing*; Taylor & Francis: London, UK, 1997; pp. 17–28.

27. Oyewole, S.A.; Haight, J.M.; Freivalds, A. The ergonomic design of classroom furniture/computer work station for first graders in the elementary school. *Int. J. Ind. Ergon.* **2010**, *40*, 437–447. [CrossRef]

28. Gouvali, M.K.; Boudolos, K. Match between school furniture dimensions and children's anthropometry. *Appl. Ergon.* **2006**, *37*, 765–773. [CrossRef]

29. García-Acosta, G.; Lange-Morales, K. Definition of sizes for the design of school furniture for Bogotá schools based on anthropometric criteria. *Ergonomics* **2007**, *50*, 1626–1642. [CrossRef]

30. Zacharkow, D. *Posture: Sitting, Standing, Chair Design, and Exercise*; Charles C Thomas Pub Limited: Springfield, IL, USA, 1988.

31. Afzan, Z.Z.; Hadi, S.A.; Shamsul, B.T.; Zailina, H.; Nada, I.; Rahmah, A.S. Mismatch between school furniture and anthropometric measures among primary school children in Mersing, Johor, Malaysia. In Proceedings of the 2012 Southeast Asian Network of Ergonomics Societies Conference (SEANES), Langkawi, Kedah, Malaysia, 9–12 July 2012; IEEE: Piscataway, NJ, USA, 2012.

32. Dianat, I.; Karimi, M.A.; Asl Hashemi, A.; Bahrampour, S. Classroom furniture and anthropometric characteristics of Iranian high school students: Proposed dimensions based on anthropometric data. *Appl. Ergon.* **2013**, *44*, 101–108. [CrossRef]

33. Bendak, S.; Al-Saleh, K.; Al-Khalidi, A. Ergonomic assessment of primary school furniture in United Arab Emirates. *Occup. Ergon.* **2013**, *11*, 85–95. [CrossRef]

34. Van Niekerk, S.-M.; Louw, Q.A.; Grimmer-Somers, K.; Harvey, J.; Hendry, K.J. The anthropometric match between high school learners of the Cape Metropole area, Western Cape, South Africa and their computer workstation at school. *Appl. Ergon.* **2013**, *44*, 366–371. [CrossRef] [PubMed]

35. Cotton, L.M.; O'Connell, D.G.; Palmer, P.P.; Rutland, M.D. Mismatch of school desks and chairs by ethnicity and grade level in middle school. *Work* **2002**, *18*, 269–280. [PubMed]

36. Parcells, C.; Stommel, M.; Hubbard, R.P. Mismatch of classroom furniture and student body dimensions: Empirical findings and health implications. *J. Adolesc. Health* **1999**, *24*, 265–273. [CrossRef]

37. Panagiotopoulou, G.; Christoulas, K.; Papanckolaou, A.; Mandroukas, K. Classroom furniture dimensions and anthropometric measures in primary school. *Appl. Ergon.* **2004**, *35*, 121–128. [CrossRef]

38. Chung, J.; Wong, T. Anthropometric evaluation for primary school furniture design. *Ergonomics* **2007**, *50*, 323–334. [CrossRef]

39. Brewer, J.M.; Davis, K.G.; Dunning, K.K.; Succop, P.A. Does ergonomic mismatch at school impact pain in school children? *Work* **2009**, *34*, 455–464. [CrossRef]

40. Jayaratne, I.L.K.; Fernando, D.N. Ergonomics related to seating arrangements in the classroom: Worst in South East Asia? The situation in Sri Lankan school children. *Work* **2009**, *34*, 409–420. [CrossRef]

41. Batistão, M.V.; Sentanin, A.C.; Moriguchi, C.S.; Hansson, G.Å.; Coury, H.J.; de Oliveira Sato, T. Furniture dimensions and postural overload for schoolchildren's head, upper back and upper limbs. *Work* **2012**, *41*, 4817–4824. [CrossRef]

42. Jayaratne, K. Inculcating the Ergonomic Culture in Developing Countries: National Healthy Schoolbag Initiative in Sri Lanka. *Hum. Factors J. Hum. Factors Ergon. Soc.* **2012**, *54*, 908–924. [CrossRef]

43. Mohamed, S.A.A.R. Incompatibility between Students' Body Measurements and School Chairs. *World Appl. Sci. J.* **2013**, *21*, 689–695.

44. Chaffin, D.B. *Occupational Biomechanics*, 3rd ed.; Andersson, G., Martin, B.J., Eds.; Wiley-Interscience Publication: New York, NY, USA, 1999.

45. Lee, Y.; Kim, Y.M.; Lee, J.H.; Yun, M.H. Anthropometric mismatch between furniture height and anthropometric measurement: A case study of Korean primary schools. *Int. J. Ind. Ergon.* **2018**, *68*, 260–269. [CrossRef]

46. Feige, U. A Threshold of ln n for Approximating Set Cover. In Proceedings of the Twenty-Eighth Annual ACM Symposium on Theory of Computing, Philadelphia, PA, USA, 22–24 May 1996; Volume 28, pp. 314–318. Available online: https://dl.acm.org/doi/abs/10.1145/237814.237977 (accessed on 20 November 2019).

47. ISO9241-5. *Ergonomic Requirements for Office Work with Visual Display Terminals (VDTs)—Part 5: Workstation Layout and Postural Requirements*; International Organization for Standardization: Geneva, Switzerland, 1998.

48. Max Roser, C.A.; Hannah, R. *Human Height*; Our World in Data: New York, NY, USA, 2013.

49. Castellucci, H.I.; Arezes, P.M.; Molenbroek, J.F.M. Analysis of the most relevant anthropometric dimensions for school furniture selection based on a study with students from one Chilean region. *Appl. Ergon.* **2015**, *46*, 201–211. [CrossRef] [PubMed]

50. Carneiro, V.; Gomes, Â.; Rangel, B. Proposal for a universal measurement system for school chairs and desks for children from 6 to 10 years old. *Appl. Ergon.* **2017**, *58*, 372–385. [CrossRef]

Article

A Part Consolidation Design Method for Additive Manufacturing based on Product Disassembly Complexity

Samyeon Kim [1] **and Seung Ki Moon** [2,*]

[1] Digital Manufacturing and Design Centre, Singapore University of Technology and Design, Singapore 487372, Singapore; samyeon_kim@sutd.edu.sg
[2] Singapore Centre for 3D Printing, School of Mechanical & Aerospace Engineering, Nanyang Technological University, Singapore 639798, Singapore
*   Correspondence: skmoon@ntu.edu.sg; Fax: +65-6792-4062

Received: 31 December 2019; Accepted: 4 February 2020; Published: 6 February 2020

Abstract: Parts with complex geometry have been divided into multiple parts due to manufacturing constraints of conventional manufacturing. However, since additive manufacturing (AM) is able to fabricate 3D objects in a layer-by-layer manner, design for AM has been researched to explore AM design benefits and alleviate manufacturing constraints of AM. To explore more AM design benefits, part consolidation has been researched for consolidating multiple parts into fewer number of parts at the manufacturing stage of product lifecycle. However, these studies have been less considered product recovery and maintenance at end-of-life stage. Consolidated parts for the manufacturing stage would not be beneficial at end-of-life stage and lead to unnecessary waste of materials during maintenance. Therefore, in this research, a design method is proposed to consolidate parts for considering maintenance and product recovery at the end-of-life stage by extending a modular identification method. Single part complexity index (SCCI) is introduced to measure part and interface complexities simultaneously. Parts with high SCCI values are grouped into modules that are candidates for part consolidation. Then the product disassembly complexity (PDC) can be used to measure disassembly complexity of a product before and after part consolidation. A case study is performed to demonstrate the usefulness of the proposed design method. The proposed method contributes to guiding how to consolidate parts for enhancing product recovery.

Keywords: additive manufacturing; complexity; modular design; part consolidation; product recovery

---

1. Introduction

Studies of product design and development have helped engineers design products systematically. Product architecture has been determined to improve manufacturability of conventional manufacturing. A part with complex geometry in the product architecture divides into multiple parts for enhancing manufacturability due to limitations of conventional manufacturing. Accordingly, design for manufacturing and assembly (DFMA) has been focused on minimizing assembly and disassembly time and cost as well as managing complexity of products by minimizing the number of parts and connectors [1–3]. Since design freedom is severely restricted by conventional design methodologies, it is difficult to achieve optimal product architecture by consolidating parts [4,5].

Additive manufacturing (AM) is revolutionizing product development by fabricating parts with complex geometry directly [6]. Design for AM (DFAM) is introduced to improve manufacturability of AM and alleviate manufacturing constraints for AM, while product lifecycle and sustainability are less considered. To explore design benefits by AM, part consolidation design methods have received attractions from designers in terms of product redesign for improving performance, but are still developing to integrate multiple parts, that are designed by limitations of conventional manufacturing, as a single part by applying AM capabilities. Accordingly, in this study, we propose a design method to consolidate parts for product recovery at the end-of-life (EOL) stage by extending conventional module identification process. Since a module consists of multiple parts, these parts in the identified module can be consolidated into a single object by AM. In the proposed method, product disassembly complexity (PDC) is used to measure difficulty while disassembling parts from a product. Therefore, the PDC plays an important role in understanding the status of product design for product recovery at the EOL stage. Since the PDC increases according to difficulty of disassembly of parts and the number of the parts and interfaces, the proposed design method aims to group parts with high disassembly difficulty into modules in order to minimize the disassembly complexity of the product at EOL stage. To assess disassembly difficulty in part level, single part complexity index (SCCI) is introduced by modifying the PDC to consider part and interface complexities simultaneously. Based on the SCCI, modules are identified by grouping parts with high SCCI value. The identified modules are considered as design boundary for part consolidation that can be fabricated by AM, so that they contribute to improving product recovery processes.

In this paper, Section 2 describes previous research and background in part consolidation and design for additive manufacturing, and then the proposed method is explained in Section 3. The proposed method described how to consolidate parts based on product disassembly complexity. Then a case study is performed with a coffee maker to demonstrate the usefulness of the proposed method in Section 4. A discussion of this study is described in Section 5. Closing remarks and future work are presented in Section 6.

2. Literature Review

Additive manufacturing (AM) process enables to produce complex parts. The AM has been evolved from rapid prototyping, which is to create a part or system rapidly as a prototype, to develop manufacturing process for creating final products directly. It alleviates design and manufacturing constraints, so that design freedom is extremely expanded [7]. In this sense, design for additive manufacturing (DFAM) has been introduced to take full advantage of the design freedom with concerning part consolidation and redesign, and hierarchical structures [6]. Most of previous studies in DFAM are to enhance performance of products while reducing costs [4,8,9], improve functional performance [10], and focus on design guidelines to print parts successfully under AM limitations [11]. Ponche, et al. [12] proposed a new DFAM methodology to consider design requirements and manufacturing specifications. The new DFAM methodology consists of three processes: part orientation and functional optimization for satisfying design requirements, and manufacturing paths optimization. Rosen [13] proposed a computer aided DFAM based on a process-structure-property-behavior framework to support part modeling, process planning, and manufacturing simulations. Thompson, et al. [4] explored design opportunities, benefits, and freedoms of AM at a part level and the macro scale, at the material level and the micro scale, and at a product level. They described part consolidation as a process to consolidate parts for assembly into a single printable object [14]. In other words, the part consolidation is considered to minimize the number of parts.

DFAM methodologies in previous studies focused on redesign of parts by using lattice structure and topology optimization. And, the redesign in module level and system level has been less addressed. According to AM capability, multi-parts can be merged as a single object instead of manufacturing and assembled parts separately and assembled. The advantages of the part consolidation are to improve manufacturing efficiency by avoiding assembly operations and reduce production cost by minimizing usage of connectors and tools for assembly [15]. There are few studies about the part consolidation. Liu [15] performed a comparative study to investigate improvement of structural performance through the part consolidation. It results in a guideline that both structural topology and build direction should be optimized to improve structural performance of consolidated parts simultaneously. Becker, et al. [16] introduced design rules for AM to help designers rethink conventional assembly design towards part consolidation. Atzeni, et al. [17] also provided design rules for AM including part consolidation. The objective of the part consolidation was to redesign parts for conventional manufacturing and minimize production costs. However, these previous studies provided general design guidelines but had less focused on how to consolidate parts into a single object. Yang, et al. [18] proposed a method of consolidating parts for AM by considering function integration to achieve better functionality and structure optimization to improve performance at a part level. Moreover, when consolidating parts by AM, sustainability should be considered. Yang, et al. [19] proposed a framework to investigate environmental impact of consolidating parts on product lifecycle. It resulted in reduction of energy consumption and environmental impact when consolidating the parts by AM. In order to focus on the end-of-life stage of product lifecycle for sustainability, it needs to be considered product lifecycle and product recovery, especially maintenance, repair, and recovery when complex parts and products approach the end-of-life stage. The product recovery is a process of restoring inherent performance of retired products. By reusing the retired product and recycling materials, companies can minimize usage of raw materials, pollution during manufacturing, and wastes at the end-of-life stage [20,21]. In addition, by replacing obsolete parts to new parts, lifespan of products can be prolonged. Accordingly, when consolidating parts by using AM processes, the product recovery should be considered to improve sustainability. To facilitate product recovery, a disassembly process is necessary to detach materials, parts, and modules from the retired products.

The disassembly process can minimize cost and time for the product recovery, and avoid damage to the quality of detached parts [22]. Therefore, previous studies of design for disassembly is mainly focused on disassembly sequence planning [2,23,24]. As complete disassembly is not cost-effective and practical, the disassembly sequence planning emphasizes on selective disassembly for product recovery and maintenance. In some studies [25,26], attributes related to the difficulty of disassembly were considered and the disassembly sequences were decided based on disassembly cost. Regarding the importance of modular design for disassembly, Ishii, et al. [27] introduced module-based design for product retirement and evaluated the compatibility of modules by calculating disassembly time and cost. Kim and Moon [28] introduced a modular design method to generate eco-modules that consider disassembly efficiency, and reusability and recyclability. In terms of manufacturing process, it is needed to assess disassembly complexity for understanding current products' conditions and then planning design strategies based on the disassembly complexity. Several papers considered process complexity with design for assembly or disassembly. ElMaraghy and Urbanic [29] introduced a product and process complexity assessment tool to understand the effects of human workers' attributes in a manufacturing line. Samy and ElMaraghy [30] proposed a product assembly complexity tool with considering handling attributes and insertion attributes during assembly operation. These assessment tools for complexity would support assembly-oriented product design and guide designers to design products with less complexity. Soh, et al. [31] measured disassembly complexity based on design for assembly and accessibility for selective disassembly operations. Limitations of these researches are that interface complexity is less considered, although the interface complexity is a major aspect of disassembly operations. Therefore, this study emphasizes on an assessment of the product disassembly complexity based on interface and component complexities simultaneously.

From the literature, three issues are identified in terms of design guidelines and sustainability. First, the design guidelines and processes for part consolidation are less considered. Most of design guidelines emphasized only on reduction of the number of parts. Second, sustainability including product recovery has rarely been considered in design for additive manufacturing. Previous studies have been researched for improving functionality through redesign. However, there are no diverse reasons for part consolidation. Finally, to support the product recovery, it is required to understand and assess disassembly complexity of a product to identify parts with high disassembly difficulty and facilitate disassembly operations. In the next section, the proposed part consolidation method to support AM is discussed in detail.

3. A Part Consolidation Design Method for Additive Manufacturing

Conventional modular design method aims to group multiple parts into modules to enhance manufacturing efficiency [32]. By shifting manufacturing paradigm from subtractive manufacturing to additive manufacturing, these multiple parts in a module can be considered as candidates for consolidation. Therefore, a part consolidation design method for AM, which is extending previous study [33,34], is proposed to group parts with high disassembly complexity into a module to enhance characteristics of products at the end-of-life (EOL) stage as shown in Figure 1. The first step is to understand function flows, such as material, signal, and energy flows, of products and physical relationships between parts. In the second step, single part complexity index (SCCI) is developed to provide information on which parts are difficult to disassemble for product recovery based on design attributes. The SCCI is an input of the third step and a modular driver for the product recovery to cluster modules from viewpoint of the EOL stage. In the third step, modules are identified based on adjacency matrix with the value of the SCCI by using Markov Cluster Algorithm. These modules would be assessed to check whether it can be manufactured by an AM technology in terms of material types. In this paper, since we focus on deciding clear design boundary for part consolidation regardless of manufacturing constraints of AM, material types are considered in this research. However, AM manufacturing constraints should be considered to determine more specific boundary for part consolidation after deciding specific AM processes. After that, parts in a module can be consolidated as a single object. It means that the concept of the module can be reinterpreted as the single part using the AM technology. Finally, to assess how product architecture with modules for part consolidation is improved to reflect product recovery, product disassembly complexity is used to compare between products with modules that is a set of parts and products with a consolidated part by AM.

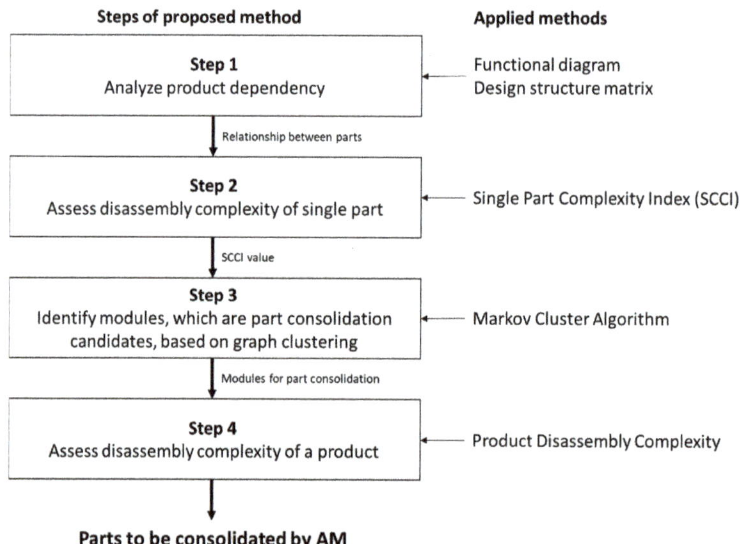

Figure 1. Overview of the proposed design method.

3.1. Product Dependency Analysis for Modular Design

Modular design has been developed to facilitate production processes, enhance product recovery including maintenance, and reduce the number of physical parts. The main principle of modular design is to improve internal coupling within modules and minimize external coupling between modules [35]. Accordingly, when the main principle of modular design is extended to the field of additive manufacturing, it would be helpful to identify parts for consolidation. This is because module identification considers functional relationships, combinability, interface standardization, and interface complexity between parts [36]. Therefore, this paper mainly focuses on identifying modules that are candidates for part consolidation with considering product recovery. To identify modules, there are many tools for the modular design: axiomatic design, functional modeling, design structure matrix, and modular function deployment [36]. In this step, a functional diagram is used to understand the function flows of a product for identifying modules as shown in Figure 2. The functional diagram consists of boxed for describing functions and three function flows: energy, material, and signal flows. Based on this information, designers can classify modules heuristically like 'Heater' to 'Water reservoir' in Figure 2. A design structure matrix (DSM) tool is applied to determine relationships between parts in a product. As shown in Figure 3 of an example of DSM, '1' represents that two parts have a relationship, while '0' represents that there is no relationship. The DSM provides fundamental information to build an adjacent matrix in Step 3.

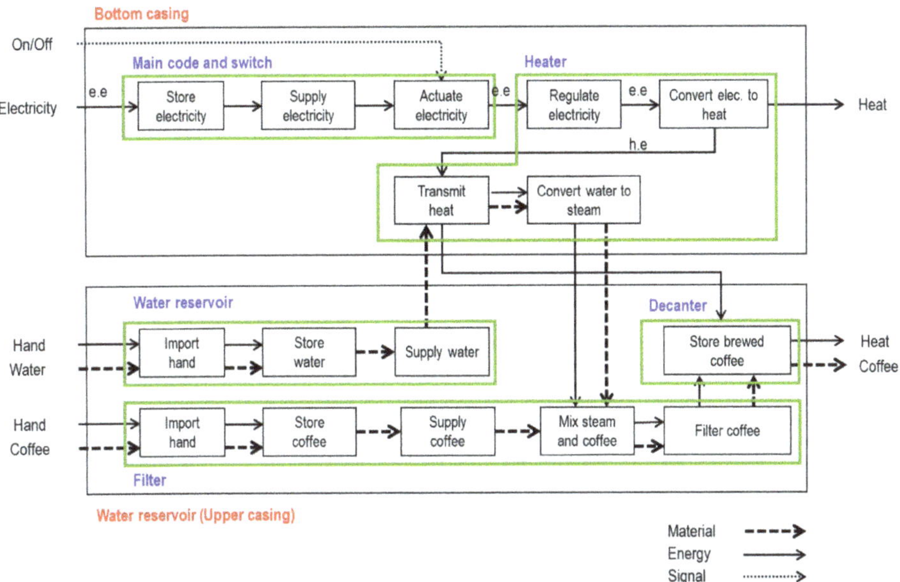

Figure 2. Functional diagram of the coffee maker.

	A	B	C	D	E
A		1	1		
B	1		1		
C	1	1		1	
D			1		1
E				1	

Figure 3. An example of design structure matrix.

3.2. Assessment of Complexity of Single Part

This research considers a 'product disassembly complexity' term as the degree of disassembly difficulty [34]. The notion of the disassembly complexity has two levels: part complexity and interface complexity. For the part complexity, it emphasizes on attributes related to handling parts: weight effect factor, size, symmetry, and grasping parts. For the interface complexity, the connector, that links parts by physical and functional relationships, such as material, energy, and signal flows, is a key attributes for manual disassembly operations. The attributes for interface are related to mechanical connector types, non-mechanical connector types, and intensity of tool use. These attributes are critical to detach parts or modules from a product.

These attributes and corresponding descriptions for parts and interfaces are described in Table 1. These attributes are converted to the disassembly difficulty factor, which is values ranging from 0 to 1. The specific values of the disassembly factor are in reference [34]. Attributes that require high disassembly difficulty are close to 1, otherwise, 0. For the part complexity, values of the disassembly attributes for a part, called as disassembly difficulty factors, are determined by measuring assembly handling time and normalizing it based on [30]. For values of the interface complexity, U-rating values are applied to measure mechanical and non-mechanical unfastening processes. The U-rating value is developed by estimating disassembly efforts based on a survey by [37] and [38]. Since the range of the U-rating value is not between 0 and 1, the U-rating value is normalized in this study.

Table 1. Disassembly attributes for manual disassembly.

Category	Attribute	Description
Part	Weight	This factor represents how difficult parts are positioned and handled according to part weight. Parts with heavy weight would need more man powers, extra tools like lift, and set-up time for parts and tools for disassembly.
	Size	A part size has an impact on both assembly and disassembly operations. When the component size is too small to grab it, it can delay the further disassembly process.
	Symmetry	The symmetry factor represents the easiness of disassembly process regarding directions for detaching parts and the difficulty of positioning parts for reassembly after disassembling the parts.
	Grasping and manipulation	Material property plays an important role in grasping parts, especially vulnerability and stiffness. Vulnerability entails damages or deformation of parts by dropping, bumping, and excessive grabbing force. Stiffness is the rigidity to resist deformation in response to an applied force, which is represented by elasticity modulus.
		As a part with low vulnerability and high stiffness can be easily grasped by a worker, the disassembly difficulty factor' value will be low. Otherwise, the disassembly difficulty factor's value is closed to 1.
Interface	Mechanical unfastening process (U-rating)	As the mechanical connectors are detachable fasteners with relevant tools, it can be recursive for assembly and disassembly. In this research, nine types of the mechanical connectors are considered as follows: screw/bolt with standard head, screw/bolt special head, nut and bolt, retaining ring/circlips, interference fit, rivets/staples, pin, cylindrical snap fit, and cantilever snap fit.
	Non-mechanical unfastening (U-rating)	The non-mechanical connectors like lead and welding material are to firmly bond components, so that disassembly can be mostly difficult.
	Tools required with low intensity/ high intensity	When using the mechanical and non-mechanical connectors, relevant tools are needed for assembly and disassembly operations. The number of tools for disassembling parts and the intensity of the tool use are considered as a disassembly attribute to represent the difficulty of disassembly.

By considering these disassembly attributes and their values, SCCI was introduced to analyze disassembly difficulty of a part by considering both part design and interface design at the same time as shown in the Equation (1) [39]. In Equation (1) for SCCI of the kth part, the weighted average value is applied to consider of part (C_k) and interface complexity indices (I_k).

$$SCCI_k = \frac{C_k \sum_1^J C_{c,j} + I_k \sum_1^N C_{i,n}}{\sum_1^J C_{c,j} + \sum_1^N C_{i,n}} \tag{1}$$

$$C_k = \frac{\sum_1^J C_{c,j}}{J} \tag{2}$$

$$I_k = \frac{\sum_1^N C_{i,n}}{N} \tag{3}$$

where, $C_{c,j}$ is a disassembly difficulty factor value of the jth attributes; $C_{i,n}$ is a disassembly difficulty factor value of nth interface attributes; C_k is the average of disassembly difficulty factors for kth part; J is the number of attributes for part complexity (here, $J = 4$); I_k is the average of disassembly difficulty factors for interfaces of kth part; and N is the number of attributes for interface complexity (here, $N = 3$) [39].

3.3. Module Identification based on Graph Clustering

In order to consider interwoven relationships between parts in a product, Markov Cluster Algorithm (MCL) is applied to group parts with high complexity into a module for AM. The MCL is used to cluster complex biological networks in the field of bioinformatics [40,41]. The MCL is a fast and scalable unsupervised clustering algorithm based on the mathematical concept of random walks.

First, an adjacent matrix, A, is developed with the value of the complexity as weight value on the edges. However, since the SCCI represents the disassembly complexity value of a single part, the SCCI value should be converted as the weight value of edges between ith part and jth part with the following equation.

$$A(i,j) = \begin{cases} w(i,j) & \textit{if ith and jth parts have relationships} \\ 0 & \textit{else} \end{cases} \tag{4}$$

$$w(i,j) = SCCI_i + SCCI_j \tag{5}$$

After building the adjacency matrix, second, Markov matrix, M, is developed to identify random walks from the adjacency matrix based on Equation (6). According to the equation, weight values in the adjacency matrix is transformed to values between 0 and 1 for representing stochastic flow from ith part to jth part.

$$M(i,j) = \frac{A(i,j)}{\sum_{k=1}^n A(k,j)} \tag{6}$$

Third, the MCL process performs two main operations: expansion and inflation. The expansion represents random walks with many steps and is the same as normal matrix multiplication. The expansion is to allow the flow to connect different regions of the graph. Nodes that have higher values with edges from a departure point to a destination point have high chance to be clustered. The inflation prunes edges with low disassembly complexity. By using Equation (7), the inflation operation makes regions with higher value on edges thicker, and makes regions with lower value on edges thinner based on the inflation parameter, r. The inflation parameter is non-negative value and used to rescale the matrix M. It results in M_{inf}, which is stochastic matrix and represents probability values of edges.

$$M_{inf}(i,j) = \frac{M(i,j)^r}{\sum_{k=1}^n M(k,j)^r} \tag{7}$$

By iterating these two main operations, parts will be grouped into modules, which is primary boundary of part consolidation for AM.

3.4. Assessment of Disassembly Complexity of a Product

Based on the aforementioned information in Table 1, the PDC can be used to represent a tendency of disassembly complexity of a product logarithmically. The total number of parts (N_c), the total number of interfaces (N_i), the number of unique parts (n_c), the number of unique interface (n_i), part complexity index (CI), and interface complexity index (II) are considered as the Equation (8) [34]. The PDC in Equation (8) is introduced by modifying the entropy theory. Accordingly, when the number of parts and interface, and values of CI and II are lower, the value of the PDC will be closed to 0.

$$PDC = \left(\frac{n_c}{N_c} + CI\right)log_2(N_c + 1) + \left(\frac{n_i}{N_i} + II\right)log_2(N_i + 1) \tag{8}$$

As shown in Equations (9) and (10), the *CI* and *II* are calculated to sum up part complexity and interface complexity of each part on Equations (2) and (3), respectively. The w_k is a weight value of the interface complexity index.

$$CI = \sum_{1}^{n_p} w_k C_k \tag{9}$$

$$II = \sum_{1}^{n_p} w_k I_k \tag{10}$$

The PDC reflects design for disassembly that recommends reduction of the number of parts. When a product has less number of parts and interfaces, the PDC will be decreased. In this study, the PDC focuses on assessing part complexity and interface complexity for a product. PDC is used to assess disassembly complexity when a product consists of modules in conventional manufacturing or consolidated parts by AM processes.

3.5. Redesign for Additive Manufacturing

Parts are designed to alleviate manufacturing constraints of conventional manufacturing and enhance assembly efficiency to minimize manufacturing cost and time. Since design paradigm is shifting from conventional manufacturing to additive manufacturing, redesign for AM is required to alleviate newly introduced manufacturing constraints and add design values by AM. To utilize the advantages of AM technologies, designers must have understanding of AM capability and limitation to ensure manufacturability of parts because they do not have experience about AM and design for AM typically [42].

Consequently, existing design methods for conventional manufacturing have been modified and improved to consider AM. Two approaches are proposed to support the modification of existing design methods [42]: (1) a partial approach and (2) a global approach. The partial approach focuses on manufacturability improvement for AM so that the results are not very far from the conventional design. Since the partial approach starts with existing design but designers have a lack of DFAM knowledge, low AM design benefits can be taken. Filippi and Cristofolini [43] and Boyard, et al. [44] combined the Design for Manufacturing (DFM) and Design for Assembly (DFA), which are conventional design methods, to apply for DFAM. Filippi and Cristofolini [43] tried to build several knowledge matrices that combine the knowledge of both design-side and manufacturing-side. Boyard, et al. [44] developed a knowledge tree for AM that indicates the inter-connection between different design stages. On the other hands, the global approach is to support exploration of AM design benefits after selecting specific AM manufacturing process characteristics while meeting the functional requirements of the parts. Therefore, topology optimization method can be utilized to take advantages of AM by resolving the stress and strain distribution on a structure. The ultimate goal of topology is saving materials [9]. Yao,

Moon, and Bi (2017) proposed an AM design feature recommendation method that can help designers organize and utilize design knowledge to explore AM-enabled design space systematically. Both partial and global approaches can guide designers to redesign existing part for adopting AM by taking AM unique capabilities. Next, we demonstrate the effectiveness of the proposed design method using a case study involving a coffee maker.

4. Case Study

To demonstrate the usefulness of the proposed design method, a case study with a coffee maker was performed. The specification of the coffee maker is described in Table 2. In the first step, the function flows of the coffee maker were described to understand functional relationship between parts for identifying modules as shown in Figure 2. Then, DSM was developed to reflect the relationships between parts in the product as shown in Table 3. In the second step, each part design and interface design between parts in the product were analyzed by using Equations (2) and (3), respectively. Based on the analyzed values, SCCI is calculated by using Equation (1) as shown in Table 4. Each value of elements in the adjacency matrix was calculated by the sum of the SCCI values of two parts based on Equations (4) and (5), so that the adjacency matrix in Table 5 is determined finally. For example, a value of the element between bottom cover (1) and bottom casing (17) was 0.020 and it was calculated by the sum of SCCI value of the bottom cover, 0.010, and SCCI value of the bottom casing, 0.010.

In the third step, MCL was applied to determine modules for product recovery, which is a design boundary for part consolidation for AM as well, by using the adjacency matrix. Since MCL is an unsupervised learning algorithm, the number of modules is determined randomly. In this case study, the number of modules converges to 7 as shown in Table 6.

Table 2. Specification of the coffee maker.

No.	Part Name	Material Type	Coffee Maker
1	Bottom cover	PP	
2	Silicon ring	Silicon	
3	Hot plate	Al	
4	Casing for heater	PP	
5	Heater	Al	
6	Power cord	Copper	
7	Water tube set	PP	
8	Silicon tube	Silicon	
9	Water reservoir	PP	
10	Steam sprout	PP	
11	Filter basket	PP	
12	Filter frame	PP	
13	Filter net	PP	4~6 cups/0.6 L
14	Filter handle	PP	Brewing time <10 min
15	Lid of coffee maker	PP	
16	Decanter	Glass	
17	Bottom casing	PP	

Table 3. Design structure matrix of the coffee maker.

DSM		1	2	3	4	5	6	7	8	9	10	11	12	13	14	15	16	17
1	Bottom cover				1													1
2	Silicon ring			1	1													
3	Hot plate		1		1	1											1	1
4	Casing for heater	1	1	1		1												1
5	Heater			1	1		1		1								1	1
6	Power cord				1													1
7	Water tube set								1									
8	Silicon tube					1		1		1	1							
9	Water reservoir								1		1	1				1		1
10	Steam sprout								1	1		1				1		
11	Filter basket									1	1		1				1	
12	Filter frame											1		1	1			
13	Filter net												1					
14	Filter handle												1					
15	Lid of coffee maker									1	1							1
16	Decanter		1		1							1						1
17	Bottom casing	1		1	1	1				1						1	1	

Table 4. Complexity information of the coffee maker.

No.	Part Name	No.	J	C_k	N	I_k	$SCCI_k$
1	Bottom cover	1	4	0.748	3	0.217	0.010
2	Silicon ring	1	4	0.828	3	0.327	0.013
3	Hot plate	1	4	0.748	3	0.217	0.010
4	Casing for heater	1	4	0.713	3	0.217	0.009
5	Heater	1	4	0.748	3	0.150	0.009
6	Power cord	1	4	0.748	3	0.483	0.014
7	Water tube set	1	4	0.748	3	0.150	0.009
8	Silicon tube	4	4	0.713	3	0.150	0.008
9	Water reservoir	1	4	0.788	3	0.110	0.010
10	Steam sprout	1	4	0.713	3	0.150	0.008
11	Filter basket	1	4	0.748	3	0.033	0.009
12	Filter frame	1	4	0.713	3	0.033	0.008
13	Filter net	1	4	0.788	3	0.277	0.011
14	Filter handle	1	4	0.748	3	0.110	0.009
15	Lid of coffee maker	1	4	0.788	3	0.133	0.010
16	Decanter	1	4	0.713	3	0.217	0.009
17	Bottom casing	1	4	0.748	3	0.217	0.010

In order to improve design feasibility of modules when adopting AM, manufacturing constraints of AM should be considered. Accordingly, total size of the module should be less than build chamber size of selected AM process and material types of parts in the module are identical except for using multi-material AM process. Furthermore, design rules for AM should be considered to improve manufacturability of product design. The design rules are mostly related to minimum thickness and overhang features that require support structure [45], which are derived from a combination of material and AM processes [46]. Therefore, designers should understand these various design rules.

In this study, we used the material type for assessing design feasibility of modules because the material type was critical when parts in a module were consolidated as a single part by sharing the same additive manufacturing processes. Accordingly, parts in modules 5 and 6 as shown in Figure 4 can be consolidated by using AM, which is 9′ and 11′ in Table 6. Accordingly, designers can consolidate parts in the modules 5 and 6 as a single part by using AM.

In the fourth step, the product disassembly complexity was applied to understand difficulty of disassembly and compare the difficulty of disassembly between a product with conventional modules and a product with consolidated parts in the modules 5 and 6. As a result, the product with consolidated parts had a lower value of the PDC than the value of PDC of the product with conventional modules as shown in Table 7, which is around 19% PDC reduction by part consolidation.

Table 5. Adjacency matrix for the single part complexity index (SCCI) of the coffee maker.

SCCI		1	2	3	4	5	6	7	8	9	10	11	12	13	14	15	16	17
1	Bottom cover				0.019													0.020
2	Silicon ring			0.023	0.022													
3	Hot plate		0.023		0.019	0.019											0.019	0.020
4	Casing for heater	0.019	0.022	0.019		0.018												0.019
5	Heater			0.019	0.018		0.023		0.018								0.018	0.019
6	Power cord					0.023												0.023
7	Water tube set								0.018									
8	Silicon tube					0.018		0.018		0.018	0.017							
9	Water reservoir								0.018		0.018	0.019				0.020		0.020
10	Steam sprout								0.017	0.018		0.017				0.019		
11	Filter basket									0.019	0.017		0.017				0.018	
12	Filter frame											0.017		0.019	0.017			
13	Filter net												0.019					
14	Filter handle												0.017					
15	Lid of coffee maker									0.020	0.019							0.020
16	Decanter			0.019		0.018						0.018						0.019
17	Bottom casing	0.020	0.020	0.019	0.019	0.023				0.020						0.020	0.019	

Table 6. Module identification and assessment.

Module No.	A Product with Conventional Modules	A Product with Parts from AM	Assessment of Modules Material Type
1	2, 3	2, 3	X
2	4	4	-
3	5	5	-
4	7, 8	7, 8	X
5	9, 10, 15	9′	O
6	11, 12, 13, 14	11′	O
7	1, 6, 16, 17	1, 6, 16, 17	X

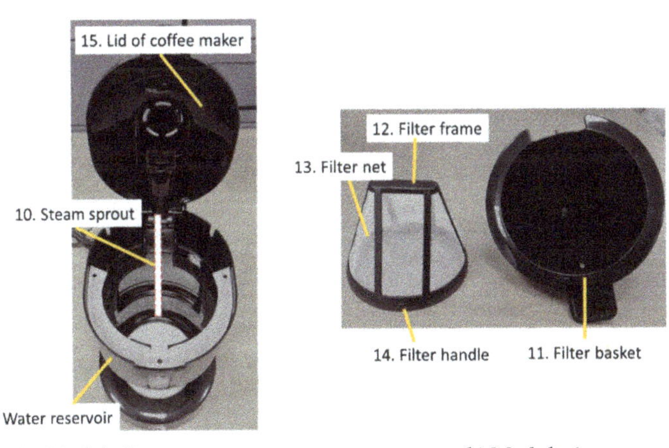

(a) Module 5   (b) Module 6

Figure 4. Parts in selected modules for part consolidation.

Table 7. Comparison of product disassembly complexity (PDC) when considering modules and parts consolidation.

Index	A Product with Conventional Modules	A Product with Parts from AM
N_c	20	15
n_c	17	12
N_i	8	8
n_i	3	3
PDC	8.765	7.079

5. Discussion

Design for AM has mainly focused on creating parts with complex geometry for improving functionality, designing parts with considering constraints of AM processes, and consolidating parts for minimizing the number of parts. To consider product recovery including maintenance, part consolidation should be planned to achieve selective disassembly. Therefore, we proposed a design method to guide how to consolidate parts by removing assembly joints that are difficult to disassemble at the EOL stage. The proposed method results in modules based on the SCCI as a modular driver, and functional and physical relationships from a functional diagram and DSM.

After identifying these modules, it is required to check whether parts can be consolidated regarding material types of the parts. Since the parts in modules 1, 4, and 7 are made of different materials like aluminum, silicon, plastic, and glass, they cannot be consolidated due to limitations of AM processes that mostly support single material. On the other hands, modules 5 and 6 contain parts that have the same material and are closed to each other physically and functionally. Furthermore, since these parts are grouped into modules because they have high SCCI values, modules 5 and 6 are appropriate candidates for part consolidation to reduce the part count of a product, which is a primary goal of part consolidation. Modules 5 and 6 will be fabricated by AM, while other modules will be manufactured by conventional manufacturing. Accordingly, the result of the proposed design method can be used as design strategy to manage which parts will be fabricated by AM selectively to support flexible manufacturing by facilitating both conventional and additive manufacturing.

However, when designers consider a design feasibility factor as maintenance frequency of the parts instead of the material type between parts in the module, consolidating the filter basket and filter consisting of filter frame, filter net, and filter handle in module 6 may be not acceptable decision because the filter should be frequently cleaned after use. Furthermore, the proposed design method can be applied to generate new candidates for part consolidation, which are parts in modules, by considering other modular drivers related to repairability, reliability, or financial benefit. These modular drivers can be represented by characteristics of parts like SCCI and characteristics between parts. For example, remained useful lifespan (RUL) of each part can be modular drivers, and then parts with the same RUL can be grouped into a module by the proposed design method with using RUL of parts instead of SCCI. Since RUL of the parts is the same, maintenance frequency would be the same. Accordingly, parts with similar lifespan can be consolidated by AM. Furthermore, feasibility analysis for selected candidates for AM should be required to identify AM benefits in terms of redesign cost, manufacturing cost and time, financial benefit, and performance enhancement against subtractive manufacturing.

6. Closing Remarks and Future Work

AM enables fabricating parts with complex geometries and consolidating multiple parts for conventional manufacturing to enhance performance by using less material and energy, compared to subtractive manufacturing. However, design for AM has mainly focused on manufacturing stage in the product lifecycle rather than end-of-life (EOL) stage. Therefore, this study considers maintenance and product recovery at the EOL stage in order to prolong product lifecycle. Since disassembly operations are closely related to efficiency of reusability and recyclability in the EOL stage, we introduced the modular design method for consolidating multiple parts to less number of parts or a single part. The disassembly complexity of each part is assessed by SCCI and then parts with high disassembly complexity are grouped into modules, which are candidates for part consolidation by using AM. Therefore, this study contributes to reduction of disassembly complexity of a product after the part consolidation.

A limitation of this study is to consider disassembly complexity for determining primary design boundary for part consolidation, which is the module. Accordingly, the proposed design method can be a starting point of product redesign for AM. As future work, other factors for product lifecycle, such as design cost, reliability of parts, maintenance requirements, and specific manufacturing constraints, will be considered to provide specific candidates for part consolidation within modules and between modules. After selecting these candidates, design feasibility of these candidates will be performed with various case studies with parts that have complex geometries after the part consolidation.

Author Contributions: Conceptualization, S.K. and S.K.M.; methodology, S.K.; formal analysis, S.K.; writing—original draft preparation, S.K.; writing—review and editing, S.K. and S.K.M.; supervision, S.K.M.; project administration, S.K.M.; funding acquisition, S.K.M. All authors have read and agreed to the published version of the manuscript.

Funding: This research was supported by the Singapore Centre for 3D Printing (SC3DP), the National Research Foundation, Prime Minister's Office, Singapore under its Medium-Sized Centre funding scheme.

Conflicts of Interest: The authors declare no conflict of interest.

References

1. Kim, S.; Baek, J.W.; Moon, S.K.; Jeon, S.M. A New Approach for Product Design by Integrating Assembly and Disassembly Sequence Structure Planning. In Proceedings of the 18th Asia Pacific Symposium on Intelligent and Evolutionary Systems (IES 2014), Singapore, 10–12 November 2014; Springer: Cham, Switzerland, 2015; Volume 1, pp. 247–257.
2. Lambert, A.J.D. Disassembly sequencing: A survey. *Int. J. Prod. Res.* **2003**, *41*, 3721–3759. [CrossRef]
3. Asikoglu, O.; Simpson, T.W. A new method for evaluating design dependencies in product architectures. In Proceedings of the 12th AIAA Aviation Technology, Integration, and Operations (ATIO) Conference and 14th AIAA/ISSMO Multidisciplinary Analysis and Optimization Conference, Indianapolis, IN, USA, 17–19 September 2012.
4. Thompson, M.K.; Moroni, G.; Vaneker, T.; Fadel, G.; Campbell, R.I.; Gibson, I.; Bernard, A.; Schulz, J.; Graf, P.; Ahuja, B.; et al. Design for Additive Manufacturing: Trends, opportunities, considerations, and constraints. *CIRP Ann.-Manuf. Technol.* **2016**, *65*, 737–760. [CrossRef]
5. Gao, W.; Zhang, Y.; Ramanujan, D.; Ramani, K.; Chen, Y.; Williams, C.B.; Wang, C.C.L.; Shin, Y.C.; Zhang, S.; Zavattieri, P.D. The status, challenges, and future of additive manufacturing in engineering. *Comput.-Aided Des.* **2015**, *69*, 65–89. [CrossRef]
6. Gibson, I.; Rosen, D.W.; Stucker, B. *Additive Manufacturing Technologies: 3D Printing, Rapid Prototyping, and Direct Digital Manufacturing*, 2nd ed.; Springer: New York, NY, USA, 2015. [CrossRef]
7. Yang, S.; Zhao, Y.F. Additive manufacturing-enabled design theory and methodology: A critical review. *Int. J. Adv. Manuf. Technol.* **2015**, *80*, 327–342. [CrossRef]
8. Yao, X.; Moon, S.K.; Bi, G. A Cost-Driven Design Methodology for Additive Manufactured Variable Platforms in Product Families. *J. Mech. Des.* **2016**, *138*. [CrossRef]
9. Lei, N.; Yao, X.; Moon, S.K.; Bi, G. An additive manufacturing process model for product family design. *J. Eng. Des.* **2016**, *27*, 751–767. [CrossRef]
10. Tang, Y.; Zhao, Y.F. A survey of the design methods for additive manufacturing to improve functional performance. *Rapid Prototyp. J.* **2016**, *22*, 569–590. [CrossRef]
11. Dinar, M.; Rosen, D.W. A Design for Additive Manufacturing Ontology. *J. Comput. Inf. Sci. Eng.* **2017**, *17*. [CrossRef]
12. Ponche, R.; Kerbrat, O.; Mognol, P.; Hascoet, J.-Y. A novel methodology of design for Additive Manufacturing applied to Additive Laser Manufacturing process. *Robot. Comput.-Integr. Manuf.* **2014**, *30*, 389–398. [CrossRef]
13. Rosen, D.W. Computer-Aided Design for Additive Manufacturing of Cellular Structures. *Comput.-Aided Des. Appl.* **2007**, *4*, 585–594. [CrossRef]

14. Wohlers, T.; Caffrey, T. *Wohlers Report 2015: 3D Printing and Additive Manufacturing State of the Industry Annual Worldwide Progress Report*; Wohlers Associates: Fort Collins, CO, USA, 2015.
15. Liu, J. Guidelines for AM part consolidation. *Virtual Phys. Prototyp.* **2016**, *11*, 133–141. [CrossRef]
16. Becker, R.; Grzesiak, A.; Henning, A. Rethink assembly design. *Assem. Autom.* **2005**, *25*, 262–266. [CrossRef]
17. Atzeni, E.; Iuliano, L.; Minetola, P.; Salmi, A. Redesign and cost estimation of rapid manufactured plastic parts. *Rapid Prototyp. J.* **2010**, *16*, 308–317. [CrossRef]
18. Yang, S.; Tang, Y.; Zhao, Y.F. A new part consolidation method to embrace the design freedom of additive manufacturing. *J. Manuf. Process.* **2015**, *20*, 444–449. [CrossRef]
19. Yang, S.; Talekar, T.; Sulthan, M.A.; Zhao, Y.F. A Generic Sustainability Assessment Model towards Consolidated Parts Fabricated by Additive Manufacturing Process. *Procedia Manuf.* **2017**, *10*, 831–844. [CrossRef]
20. Kwak, M.; Kim, H.M. Evaluating End-of-Life Recovery Profit by a Simultaneous Consideration of Product Design and Recovery Network Design. *J. Mech. Des.* **2010**, *132*. [CrossRef]
21. Thierry, M.; Salomon, M.; van Nunen, J.; van Wassenhove, L. Strategic Issues in Product Recovery Management. *Calif. Manag. Rev.* **1995**, *37*, 114–135. [CrossRef]
22. Smith, S.; Smith, G.; Chen, W.-H. Disassembly sequence structure graphs: An optimal approach for multiple-target selective disassembly sequence planning. *Adv. Eng. Inform.* **2012**, *26*, 306–316. [CrossRef]
23. Lambert, A.J.D. Linear programming in disassembly/clustering sequence generation. *Comput. Ind. Eng.* **1999**, *36*, 723–738. [CrossRef]
24. Behdad, S.; Berg, L.P.; Thurston, D.; Vance, J. Leveraging virtual reality experiences with mixed-integer nonlinear programming visualization of disassembly sequence planning under uncertainty. *J. Mech. Des.* **2014**, *136*. [CrossRef]
25. Tseng, Y.-J.; Yu, F.-Y.; Huang, F.-Y. A green assembly sequence planning model with a closed-loop assembly and disassembly sequence planning using a particle swarm optimization method. *Int. J. Adv. Manuf. Technol.* **2011**, *57*, 1183–1197. [CrossRef]
26. Rickli, J.L.; Camelio, J.A. Multi-objective partial disassembly optimization based on sequence feasibility. *J. Manuf. Syst.* **2013**, *32*, 281–293. [CrossRef]
27. Ishii, K.; Eubanks, C.F.; Di Marco, P. Design for product retirement and material life-cycle. *Mater. Des.* **1994**, *15*, 225–233. [CrossRef]
28. Kim, S.; Moon, S.K. Assessing and Generating Modules for Product Recovery. In Proceedings of the ASME 2015 International Design Engineering Technical Conferences & Computers and Information in Engineering Conference (DETC/CIE2015), Boston, MA, USA, 2–5 August 2015.
29. ElMaraghy, W.H.; Urbanic, R.J. Modelling of Manufacturing Systems Complexity. *CIRP Ann.-Manuf. Technol.* **2003**, *52*, 363–366. [CrossRef]
30. Samy, S.N.; ElMaraghy, H. A model for measuring products assembly complexity. *Int. J. Comput. Integr. Manuf.* **2010**, *23*, 1015–1027. [CrossRef]
31. Soh, S.L.; Ong, S.K.; Nee, A.Y.C. Application of Design for Disassembly from Remanufacturing Perspective. *Procedia CIRP* **2015**, *26*, 577–582. [CrossRef]
32. Otto, K.; Hölttä-Otto, K.; Simpson, T.W.; Krause, D.; Ripperda, S.; Moon, S.K. Global Views on Modular Design Research: Linking Alternative Methods to Support Modular Product Family Concept Development. *J. Mech. Des.* **2016**, *138*. [CrossRef]
33. Kim, S.; Moon, S.K. Disassembly Complexity-Driven Module Identification for Additive Manufacturing. In Proceedings of the 24th ISPE Inc. Transdisciplinary Engineering, Singapore, 10–14 July 2017.
34. Kim, S.; Moon, S.K.; Jeon, S.M.; Oh, H.S. A disassembly complexity assessment method for sustainable product design. In Proceedings of the 2016 IEEE International Conference on Industrial Engineering and Engineering Management (IEEM), Bali, Indonesia, 4–7 December 2016; pp. 1468–1472.
35. Ulrich, K.T. The role of product architecture in the manufacturing firm. *Res. Policy* **1995**, *24*, 419–440. [CrossRef]
36. Stjepandić, J.; Ostrosi, E.; Fougères, A.-J.; Kurth, M. Modularity and Supporting Tools and Methods. In *Concurrent Engineering in the 21st Century: Foundations, Developments and Challenges*; Stjepandić, J., Wognum, N., J.C. Verhagen, W., Eds.; Springer International Publishing: Cham, Switzerland, 2015; pp. 389–420. [CrossRef]
37. Das, S.K.; Naik, S. Process planning for product disassembly. *Int. J. Prod. Res.* **2002**, *40*, 1335–1355. [CrossRef]

38. Das, S.K.; Yedlarajiah, P.; Narendra, R. An approach for estimating the end-of-life product disassembly effort and cost. *Int. J. Prod. Res.* **2000**, *38*, 657–673. [CrossRef]
39. Kim, S. Sustainable Product Family Design and a Platform Strategy. Ph.D. Thesis, Nanyang Technological University, Singapore, 2017.
40. Lei, X.; Wang, F.; Wu, F.-X.; Zhang, A.; Pedrycz, W. Protein complex identification through Markov clustering with firefly algorithm on dynamic protein–protein interaction networks. *Inf. Sci.* **2016**, *329*, 303–316. [CrossRef]
41. Kim, S.; Moon, S.K. Eco-modular product architecture identification and assessment for product recovery. *J. Intell. Manuf.* **2016**. [CrossRef]
42. Ponche, R.; Hascoet, J.Y.; Kerbrat, O.; Mognol, P. A new global approach to design for additive manufacturing. *Virtual Phys. Prototyp.* **2012**, *7*, 93–105. [CrossRef]
43. Filippi, S.; Cristofolini, I. The Design Guidelines (DGLs), a knowledge-based system for industrial design developed accordingly to ISO-GPS (Geometrical Product Specifications) concepts. *Res. Eng. Des.* **2007**, *18*, 1–19. [CrossRef]
44. Boyard, N.; Rivette, M.; Christmann, O.; Richir, S. A design methodology for parts using additive manufacturing. In Proceedings of the International Conference on Advanced Research in Virtual and Rapid Prototyping, Leiria, Portugal, 1–5 October 2013.
45. Kranz, J.; Herzog, D.; Emmelmann, C. Design guidelines for laser additive manufacturing of lightweight structures in TiAl6V4. *J. Laser App.* **2015**, *27*, S14001. [CrossRef]
46. Kim, S.; Rosen, D.W.; Witherell, P.; Ko, H. A Design for Additive Manufacturing Ontology to Support Manufacturability Analysis. *J. Comput. Inf. Sci. Eng.* **2019**, *19*. [CrossRef]

MDPI

St. Alban-Anlage 66

4052 Basel

Switzerland

Tel. +41 61 683 77 34

Fax +41 61 302 89 18

www.mdpi.com

Applied Sciences Editorial Office

E-mail: applsci@mdpi.com

www.mdpi.com/journal/applsci